MATH
GAMES
with Bad Drawings

OTHER BOOKS BY BEN ORLIN

Math with Bad Drawings

Change Is the Only Constant

MATH GAMES

with Bad Drawings

75¼ SIMPLE, CHALLENGING, GO-ANYWHERE GAMES— AND WHY THEY MATTER

BEN ORLIN

BLACK DOG
& LEVENTHAL
PUBLISHERS
NEW YORK

Black Dog & Leventhal Publishers
Hachette Book Group
1290 Avenue of the Americas
New York, NY 10104

www.hachettebookgroup.com
www.blackdogandleventhal.com

First edition: April 2022

On page 165, lyrics to Jonathan Coulton's "Skullcrusher Mountain" are used by permission.

Black Dog & Leventhal Publishers is an imprint of Perseus Books, LLC, a subsidiary of Hachette Book Group, Inc. The Black Dog & Leventhal Publishers name and logo are trademarks of Hachette Book Group, Inc.

The publisher is not responsible for websites (or their content) that are not owned by the publisher.

The Hachette Speakers Bureau provides a wide range of authors for speaking events. To find out more, go to www.HachetteSpeakersBureau.com or call (866) 376-6591.

Print book interior design by Headcase Design

Library of Congress Cataloging-in-Publication Data

Names: Orlin, Ben, author.
Title: Math games with bad drawings : 74 1/2 simple, challenging,
    go-anywhere games-and why they matter / Ben Orlin.
Description: First edition. | New York, NY : Black Dog & Leventhal, 2022. |
    Includes bibliographical references and index. | Summary: "Best-selling
    author and worst-drawing artist Ben Orlin expands his oeuvre with this
    interactive collection of mathematical games. Each taking a minute to
    learn and a lifetime to master, this treasure chest of 70-plus games
    will delight, educate, and entertain"—Provided by publisher.
Identifiers: LCCN 2021030732 (print) | LCCN 2021030733 (ebook) | ISBN
    9780762499861 (hardcover) | ISBN 9780762499854 (ebook)
Subjects: LCSH: Mathematics—Popular works. | CYAC: Mathematical
    recreations.
Classification: LCC QA93 .O865 2022 (print) | LCC QA93 (ebook) | DDC
    793.74—dc23
LC record available at https://lccn.loc.gov/2021030732
LC ebook record available at https://lccn.loc.gov/2021030733

ISBNs: 978-0-7624-9986-1 (hardcover); 978-0-7624-9985-4 (ebook)

Printed in China

1010

10 9 8 7 6 5 4 3 2

For Casey,
who teaches me new games every day,
all of them magical
and many of them deeply confusing

CONTENTS

INTRODUCTION

LET'S BEGIN WITH a riddle. What, exactly, distinguishes you from a chimpanzee?

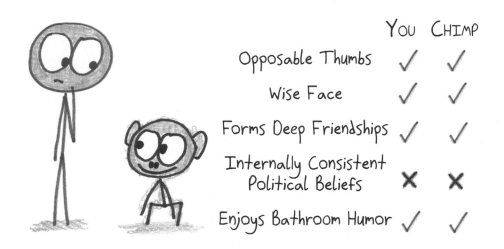

	You	Chimp
Opposable Thumbs	✓	✓
Wise Face	✓	✓
Forms Deep Friendships	✓	✓
Internally Consistent Political Beliefs	✗	✗
Enjoys Bathroom Humor	✓	✓

Answer: The chimpanzee began as a baby chimp, then grew up, whereas you began as a baby chimp, then stayed that way.

Seriously, look at yourself: furless skin, tiny jaw, enormous rounded cranium—these are traits that our ape cousins lose as they age, yet you have stubbornly kept. No judgment; I've done it too. We humans retain childlike traits into adulthood, clinging to what biologist Stephen Jay Gould called "eternal youth." The technical term is *neoteny*, and among primates, it's kind of our calling card. The best part is that we don't just look like baby chimps. We also act like them: mimicking, exploring, puzzling—in a word, *playing*.

That, my baby-faced friends, is how we became the geniuses of the primate order. That's how we built our pyramids, left our footprints on the moon, and recorded our multiplatinum album *Abbey Road*. Not by outgrowing foolishness, but by refusing to. The secret to our brilliance is that we never stop learning, and the secret to our learning is that we never stop playing.

So, let's play, shall we?

HOW TO PLAY THIS BOOK

What do you need?

1. **A few common household items.** I've strived for games that require only pen and paper, though for some, you'll need a bit more. The details are spelled out in each chapter, and summarized in tables in the Conclusion. (Note that dice can easily be simulated; search "roll dice" online.)

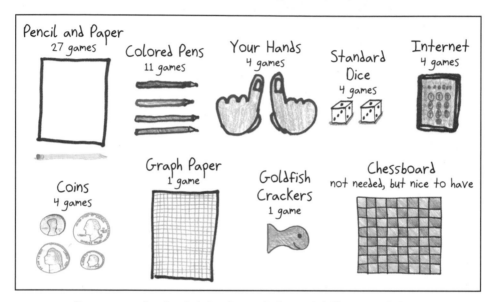

Games counted under their hardest-to-find material. VARIATIONS & RELATED GAMES not included. Offer void where prohibited by law. Though if you're someplace where the laws prohibit games, you've got way bigger problems.

2. **Playmates.** Plenty of math books are for solo play. Not this one. I wrote it during a year of pandemic-induced "social distancing," and not surprisingly, the result is a love letter to social togetherness. Thus, except for a few one-player games, you'll need companions. Also, although I've written this book for an audience of old baby chimps like myself, a 10-year-old could enjoy almost every game, and many are suitable for those as young as six.

3. **A healthy dose of neoteny.** "Many animals display flexibility and play in childhood," wrote Stephen Jay Gould, "but follow rigidly programmed patterns as adults." As a math teacher, I admit that our math lessons often seem designed for some other animal, one of those rigid pattern-followers. Perhaps termites. No surprise that such lessons capture our thinking at its worst: paralyzed, plodding, anxious. For this book, forget all that and summon your true nature, your inner baby.

What's the goal? To bring out the best in human thought.

What are the rules?

1. This book tackles a specific and uniquely human kind of play: **games**, also known as "**play with rules**." They range from those with myriad rules (like Monopoly) to those with just one (like "the floor is lava"), from occasions of merciless competition (like Monopoly) to ones of profound collaboration (like "the floor is lava"), from the worst artifacts of human culture (like Monopoly) to the best (like "the floor is lava").

    For this book, I have sought games with **simple, elegant rules** that give rise to **rich, complex play**. You know the saying: "a minute to learn, a lifetime to master."

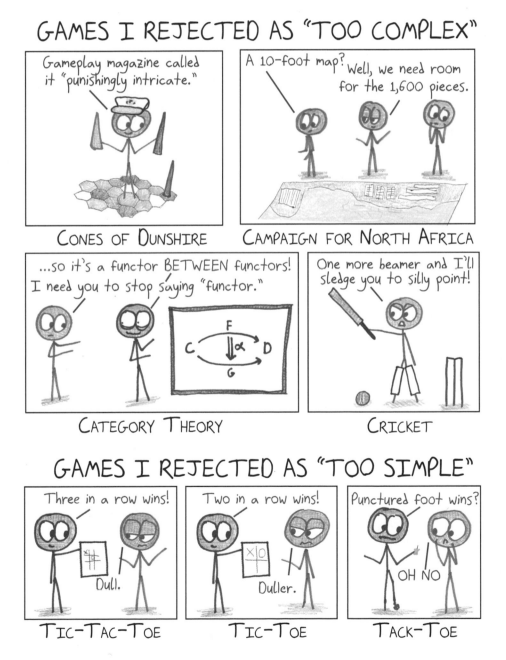

GAMES I REJECTED AS "TOO COMPLEX"

CONES OF DUNSHIRE   CAMPAIGN FOR NORTH AFRICA

CATEGORY THEORY   CRICKET

GAMES I REJECTED AS "TOO SIMPLE"

TIC-TAC-TOE   TIC-TOE   TACK-TOE

2. **What do you mean by "mathematical games?"** Good question. I first fielded it from Vito Sauro, one of Minnesota's friendliest experts in tabletop gaming. Almost every board game, he pointed out, consists of a thematic skin over a mathematical skeleton. Would my book attempt to cover all games that have ever existed?

No, no, I told Vito. A mathematical game, by my definition, is one that makes you go *Mmm-mmm, that's mathy.*

Vito considered this (1) a total non-answer, and (2) fairly satisfactory. In any case, I've tried to compile **timeless games of logic, strategy, and spatial reasoning**. My three criteria were: (1) fun, (2) easy to play, and (3) mathematically thought-provoking.[1]

3. The book has five sections: **Spatial Games**, **Number Games**, **Combination Games**, **Games of Risk and Reward**, and **Information Games**. There's an element of whimsy to these classifications. Don't think of them as a perfect taxonomy, with each specimen filed away. They're more a kind of mood lighting, highlighting distinctive features. For example, chess could comfortably belong to any of the five categories—though it'd look a little different under each light.

"Each piece projects force... rays of light and darkness extending across the board."
—Greg Costikyan

SPATIAL

"One bad move nullifies forty good ones."
—Bernhard Horwitz

NUMBER

"Play the opening like a book, the middlegame like a magician, and the endgame like a machine."
—Rudolf Spielmann

COMBINATION

"To avoid losing a piece, many a person has lost the game."
—Savielly Tartakower

RISK AND REWARD

"If winning, clarify; if losing, complicate."
—Bruce Pandolfini

INFORMATION

---

1 Although I'm a teacher, I've steered clear of classroom fare like "polynomial dominos," "simultaneous equation *Jeopardy!*" and "who can do their homework the fastest."

Each section begins with a playful essay on the relevant branch of mathematics. After that come **five featured games**, roughly increasing in complexity (though each new section is a fresh start). Last is a chapter of **miscellaneous games described in brief** (including some of my very favorites).

4. Each featured game follows the same structure. First, in **How to Play**, I'll lay out the mechanics, including **what you need**, the **goal**, and the **rules**.

Second, in **Tasting Notes**, I'll elaborate on the flavor of the gameplay, the elusive *je ne sais quoi*. You may glean a few strategic tips, but that's not my aim. I'm focused on teasing out the subtle hues and shades of mathematical play, a variety so exquisite that it makes wine look like sad old grape juice.[2]

Third, in **Where It Comes From**, I'll tell you what I know of the game's origins. Some are ancient and timeless, some are silly and novel, and some are both at once (don't ask me how that works, it just does).

Fourth, in **Why It Matters**, I'll tell you why this game brings out the best in human thought. Maybe it models the quantum structure of matter. Maybe it reveals the austere beauty of topology or the cold logic of gerrymandering. Maybe it unlocks your inner genius or, better yet, your inner chimpanzee. In any case, I view this as the chapter's crux, and the book's driving purpose.

---

2   Wine, for those of you too young to know, is sad old grape juice.

Finally, in VARIATIONS AND RELATED GAMES, I'll propose a few pleasant directions in which you can wander off to explore. Some are minor rule tweaks, while others are whole new games connected to the original by history, concept, or spirit.

5. In the book's Conclusion, there are exhaustive tables of the included games and a crowd-pleasing bibliography written in the form of frequently asked questions.

Oh, and I'll also explain how I arrived at the unorthodox number of 75¼ games included in this book. If you're asking, "What's the quarter game?" then please rest assured: It's *much* more complicated than that.

TASTING NOTES FOR THIS BOOK

You can, if you like, treat this as an ordinary work of nonfiction. Turn the pages. Smile politely at the jokes. Mutter to yourself, "Wow, what bad drawings. I sure am getting my money's worth." Moving chapter to chapter, front to back, game to game, you'll have a perfectly pleasant time.

You'll also miss all the fun.

This book is meant to be played. A human playing with math is like an elephant playing with its trunk, a bird with its wings, or a Batman with its fancy car. It's a creature doing what it was born to do. Your capacity for mathematical thinking is a gift of extraordinary dimensions, a force unmatched in the animal kingdom, except perhaps by a cat's capacity for contempt. Please don't leave this evolutionary present in its wrapping paper. Take it out. Play with it. Or at least, like a cat, play with the wrapping paper.

Most of the games are multiplayer. I hope you find a playmate who shares your curiosity, with whom you can prod and poke the games, analyzing as you go. "Only dead mathematics can be taught where competition prevails," said the mathematician Mary Everest Boole. "Living mathematics must always be a communal possession." The way I see it, even competitive games are collaborative projects, in which minds unite to build extraordinary chains of logic and strategy. David Bronstein called this "thinking for two." Karl Menninger called it "a progressive interpenetration of minds." I call it "play."

All that said, this *is* a book, and I do hope you read it. Each game shines light on a deep truth about mathematics, from the combinatorial explosion to information theory. Those mathematical truths, in turn, shine light on the games. If that all sounds blindingly bright, don't worry, your eyes will adjust. As the cleric Charles Caleb Colton once wrote, "The study of mathematics, like the Nile, begins in minuteness but ends in magnificence."

WHERE MATHEMATICAL GAMES COME FROM

The games in this book come from Parisian universities, Japanese schoolyards, Argentine magazines, humble hobbyists, shameless self-promoters, raucous gambling halls, inebriated scholars, and excitable young children. The games are varied because math is varied, silly because math is silly. And the games belong to everyone, because no matter what those daunting formulas and sneering gatekeepers seem to say, math belongs to everyone, too.

By and large, I've drawn games from four realms:

1. **Traditional children's pastimes**, such as Battleship, Chopsticks, and Dots and Boxes.

2. **Recreational hobby games**, such as Teeko, Paper Boxing, and Amazons.

3. **Concepts devised by mathematicians**, such as Sim, Sprouts, and Domineering.

4. **Weirdly fun classroom games**, such as Neighbors, Outrangeous, and 101 and You're Done.

How do these games emerge? What sparks a mathematical fire? Having designed nine of the games myself, I ought to know. But there's no one path, no common origin story. India gave us chess, China gave us go, Madagascar gave us fanorona, and my two-year-old nephew Skander gave us "dance around the puzzle chanting 'mowawawawa.'"

Why are mathematical games so universal? I truly don't know. But perhaps it's because the universe is so mathematical.

Case in point: In 1974, a geneticist named Marsha Jean Falco began drawing symbols on index cards. This was a research tool: Each card represented a dog, and each symbol represented a DNA sequence in that dog's genome. But as she shuffled and rearranged the cards, all of the specifics fell away. She started to see pure combinations, abstract patterns. The play of logic. The logic of play. "Matter does not engage [mathematicians'] attention," wrote Henri Poincaré, "they are interested by form alone." A veterinarian looking over Marsha's shoulder began asking questions, and before long, Marsha got an idea for a game.

Thus was born a favorite pastime of Stephen Hawking, a favorite research topic among leading mathematicians, and one of the most popular card games of the 20th century: Set.

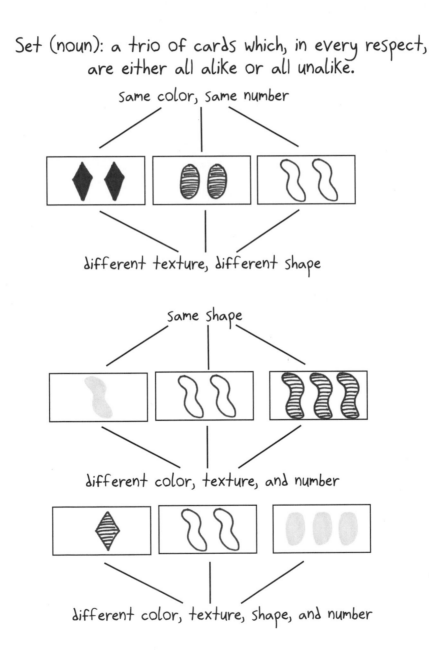

Set (noun): a trio of cards which, in every respect, are either all alike or all unalike.

Same color, same number

different texture, different shape

Same shape

different color, texture, and number

different color, texture, shape, and number

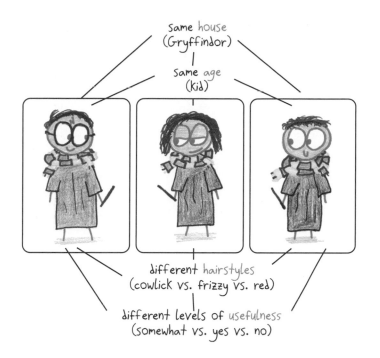

Same house
(Gryffindor)

Same age
(kid)

different hairstyles
(cowlick vs. frizzy vs. red)

different levels of usefulness
(somewhat vs. yes vs. no)

That same year, 1974, a Hungarian architect set himself a structural challenge: Could you make a big block out of smaller blocks that moved independently? He tried. He succeeded. And then, on a fateful whim, he stuck colored paper to each side and began twisting. "It was tremendously satisfying to watch this color parade," he later recalled. But eventually, "like after a nice walk when you have seen many lovely sights, I decided it was time to go home . . . [L]et us put the cubes back in order."

He tried. He failed. And, playful human that he was, he kept trying. After a month's effort, he finally restored the cube to its original state.

Thus did Ernő Rubik create the bestselling toy in human history.

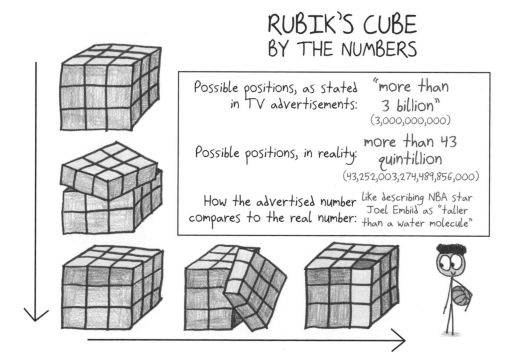

RUBIK'S CUBE
BY THE NUMBERS

Possible positions, as stated in TV advertisements: "more than 3 billion" (3,000,000,000)

Possible positions, in reality: more than 43 quintillion (43,252,003,274,489,856,000)

How the advertised number compares to the real number: like describing NBA star Joel Embiid as "taller than a water molecule"

Set and the Rubik's Cube show us the two fundamental pathways of mathematical thought. You can begin with a piece of reality, as Marsha did, and seek its abstract structure, or you can begin with an abstract structure, as Ernő did, and seek its meanings in reality. In that sense, Set and the Rubik's Cube don't just enable others to play; they are themselves the fruits of playful thinking, the idle art of genius apes who never stopped learning.

WHY MATHEMATICAL GAMES MATTER

Because they bring out the best in human thought.

In 1654, a gambler wrote to two mathematicians to share a puzzle. It goes like this: Imagine two people playing a simple 50/50 game of chance, scoring 1 point per round. The first to 7 points wins $100. But partway through, with one player leading 6 to 4, the game is interrupted. What's the fair way to split the prize?

The two mathematicians, Blaise Pascal and Pierre de Fermat, solved the problem,[3] and better yet, their solution helped to birth the mathematical study of uncertainty, known today as *probability theory*.

A fundamental tool of modernity, born from a simple puzzle about a game of chance.

---

3 Spoilers: The trailing player's only hope for victory is to win each of the next three rounds. That has a 1 in 8 chance of happening. Thus, this player deserves one eighth of the prize, or $12.50. The leader gets $87.50.

Here's another true story. On Sunday afternoons in the 1700s, the citizens of Königsberg (known today as Kaliningrad) enjoyed wandering the four regions of their riverside city, aiming to cross the seven bridges—the Blacksmith's, the Connecting, the Green, the Merchant's, the Wooden, the High, and the Honey— exactly once each. No one could manage it. Then, in 1735, the mathematician Leonhard Euler proved why: It was impossible. No such path could exist. Today, we recognize his proof as the dawn of *graph theory*, the study of networks, which underlies everything from social media to internet search algorithms to epidemiology. Google, Facebook, and the fight against COVID-19 can trace their lineage to the amusements of a Prussian afternoon.

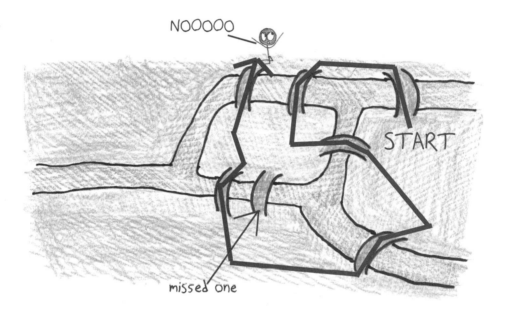

Need another example? How about John Horton Conway, a larger- than-life mathematician who passed away while I was writing this book. He explored math in all its staggering variety, from cellular automata to abstract algebra. Yet he kept returning, again and again, to games. His favorite discovery was the surreal numbers, which encoded the structure of two-player games into a numerical system. His most heralded (and thus least favorite) discovery showed how worldlike complexity could arise from a few simple rules; it was called the Game of Life.

"I was struck," writes mathematician and admirer Jim Propp, "by the way his ideas about games turned out to play a role in his work on lattices, codes, and packings . . . What are the chances that a mathematician who loved games would have the luck to find that games secretly underlie other subjects he studies?"

I could go on—a weekly poker night inspired John von Neumann to develop game theory, whose strategic insights now penetrate ecology, diplomacy, and economics—but I'm not here to worship applications. I don't really care that mathematical play has helped to mint billionaires or to create trillions of dollars of wealth. My point is that mathematical play does this *as an accidental by-product.*

When you look up from your game and realize you've inadvertently changed the course of human history, you know you're playing with a special kind of fire.

"All good thinking is play," writes Mason Hartman. She means that our best thoughts explore ideas the way a baby chimp explores the forest, with a kind of freedom and abandon. It's not a game of Parcheesi, with every move geared toward victory; rather, it's a game of make-believe, a game of "yes, and . . . ," a game passed from generation to generation, a torch that never goes dark. "A finite game is played for the purpose of winning," wrote James Carse, "an infinite game for the purpose of continuing to play."

We often see mathematics as a series of finite games. Questions to answer. Puzzles to solve. Theorems to prove. But taken together, they form a vast and never-ending game, encompassing the thoughts of every sentient ape. "I love mathematics," said mathematician Rózsa Péter, "because man has breathed his spirit of play into it, and because it has given him his greatest game—the encompassing of the infinite."

Personally, I say man's greatest game is "the floor is lava," but I get a kick out of encompassing the infinite now and then. I cordially invite you to join me.

I
SPATIAL GAMES

YOU'RE ABOUT TO meet five games, each belonging to a different kind of space. If nothing else, I hope you take away that lesson: *There are different kinds of space.*

Dots and Boxes unfolds on a tight rectangular grid, like a planned city. Sprouts unfurls across an oozing, snaking dreamscape. Ultimate Tic-Tac-Toe envisions a fractal world of microcosms, macrocosms, echoes. Dandelions is a game of windswept plains and stark vectors. Finally, Quantum Tic-Tac-Toe inhabits a space so eerie it barely feels like space at all. Put them together, and you can see why mathematicians talk not about "geometry" but "geometries," whole different ways of conceptualizing space and its contents. "One geometry cannot be more true than another," wrote mathematician Henri Poincaré, "it can only be more convenient."

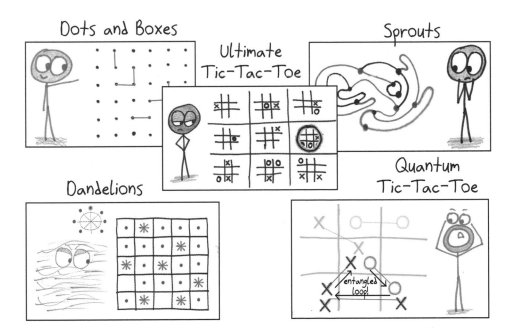

Yet these diverse games share one thing in common: They're flat. They're 2D experiences trying to shed light on a 3D world, like shadows in reverse.

It's a funny gig, being a modern human. Whereas our ancestors swung like Tarzan from tree to tree, I swing like Jane from book to book, page to page, paper to paper. I have a brain built for a 3D world of depth and motion, yet I've consigned it to a 2D world of documents and screens, thin slices of a thick reality.

Well, if we can't bring the ape back to the jungle, geometric games do the next best thing: They bring the jungle back to the ape. They elaborate flatness into depth, 2D into 3D.

I'll show you what I mean with three quick bonus games.

First bonus game: the 1979 arcade classic Asteroids, in which you maneuver an arrow-shaped spaceship around a screen. That screen is the whole universe: Fly off the edge, and you reappear on the opposite side. The resulting experience feels sphere-like, with travel in any direction leading you back to your starting location.

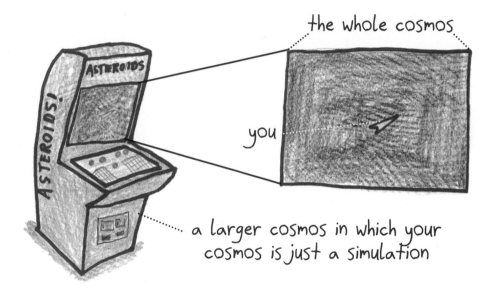

the whole cosmos

you

a larger cosmos in which your cosmos is just a simulation

Yet it's not actually a sphere. First, by connecting the screen's left and right edges, the game's designers created a kind of cylindrical world. Then, by linking the screen's top and bottom edges, they connected the two rims of this cylinder. The result is not a sphere but a doughnut shape, known to die-hard math fans as a *torus*.[1]

Asteroids inhabits a toroidal universe. Someone should tell NASA.

---

1   In his book *New Rules for Classic Games*, R. Wayne Schmittberger suggests applying Asteroids' spatial logic to Scrabble, so that a word can vanish off the bottom and continue at the top, or vanish off the right edge and continue on the left. "One of the amusing results of playing Toric Scrabble," he writes, "is board positions that look not only illegal but totally ridiculous by conventional Scrabble standards. A word fragment or a single letter will be floating along an edge, seemingly unconnected to anything, when it is actually part of a word on the opposite edge. It's a great way to worry the kibitzers." I recommend applying the same toroidal logic to other games in this book, such as Sprouts, Sequencium, and Amazons.

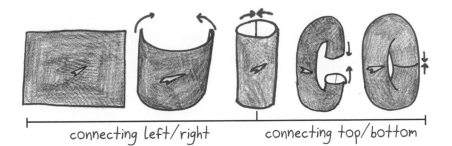

connecting left/right   connecting top/bottom

For our second 2D-to-3D game, we turn to mathematician Ingrid Daubechies. "When I was eight or nine," she once recalled, "the thing I liked best when playing with my dolls was to sew clothes on them. It was fascinating to me that by putting together flat pieces of fabric one could make something that was not flat at all, but followed curved surfaces."

Decades later, her work on wavelets would drive forward the technology of image compression. In a sense, it's the same game: flatness and curvature, solidity and surface, depth and compression.

Geometry, I argue, is nothing other than the ancient mathematics of dressing up your dolls.

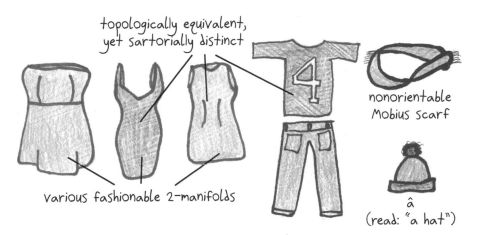

topologically equivalent, yet sartorially distinct

nonorientable Mobius scarf

various fashionable 2-manifolds

$\hat{a}$
(read: "a hat")

For our final 2D-to-3D game, I give you the artist M. C. Escher. Maybe you've seen some of his images: two hands drawing one another, a tessellation of birds and fish, an impossible staircase that leads up and up and up and up. Mathematicians love his stuff because it's just like theirs: the silly play of profound ideas. "[It's] a pleasure," he wrote, "knowingly to mix up two- and three-dimensionalities, flat and spatial, and to make fun of gravity."

"All my works are games," he liked to say. "Serious games."

To me, there's no better way to explore the different worlds of geometry than with games and puzzles. Such experiences give us, in the words of mathematician John Urschel, "a glimpse of the possible pathways of thought." They offer brief and vivid experiences of whole different realities.

You and I are apes at heart. We can't help thinking spatially. So it's a good thing that space comes in a thousand flavors and styles, each stranger and more wondrous than the last.

DOTS AND BOXES

A GAME OF SQUARES

In the introduction to his 130-page book *Dots and Boxes: Sophisticated Child's Play*, the mathematician Elwyn Berlekamp called this game "the mathematically richest popular child's game in the world." Whether he meant to call it a sophisticated game for popular children, a popular game for sophisticated children, or a sophisticated and worldly game for rich and popular children, the message is clear: This game slaps.

In this brief chapter, I can't lay out a complete theory of Dots and Boxes. Instead, I'll lay out something better: a complete theory of mathematical inquiry, straight from the scholar who first published the rules to this game.

Will reading these pages transform you into a rich, popular, and sophisticated child? Legally, I can't promise that. So just look at my winking eye, and sally forth.

HOW TO PLAY

What do you need? Two players, a pen, and an array of dots. I recommend 6-by-6, but any rectangular array works.

What's the goal? Claim more boxes than your opponent.

What are the rules?

1. Take turns drawing little vertical or horizontal lines to **connect adjacent dots**.

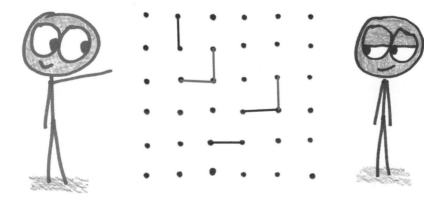

2. Whoever draws the **fourth line to complete a small box gets to claim the box as their own** (by writing their initial inside), and then **immediately takes another turn**.

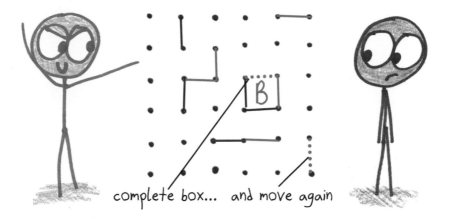

complete box... and move again

This rule may allow you to claim several boxes in a row before your opponent has a chance to move again.

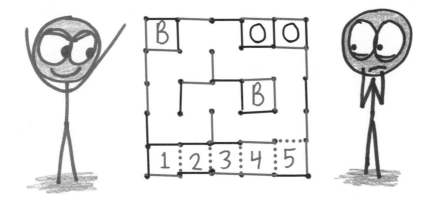

3. Play until the grid is full. **Whoever claims more boxes is the winner.**

7 squares

18 squares

TASTING NOTES

I first played this game in my childhood basement, amid shelves of VHS tapes and the rumbling footfalls of passing dinosaurs. My siblings and I lacked strategic sophistication: We moved pretty much at random, making sure not to draw the third line on any box (which would let your opponent draw the fourth), and otherwise scattering our marks willy-nilly.[2] Then, at some point, no safe moves remained. That's when things got tense.

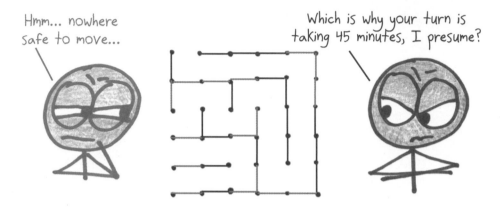

Sacrifice was now unavoidable—yet not all sacrifices were equal. Some moves might give your opponent just one or two boxes; others, practically the whole board. I always tried to give away the smallest region possible, hoping to reserve the larger ones for myself.

Years later, while working on this book, I learned a crucial stratagem. It's simple to execute, yet sufficient to beat 99% of novices: **the double cross**. The idea is that, when your opponent is set up for a triumphant next turn, you don't give it to them. Instead, cut your own turn short by skipping over the second-to-last move. You thus sacrifice two boxes, which your opponent

---

2   Sometimes, a sneaky second player would mirror the first player's moves, so that the board would look the same if rotated 180°. This guaranteed the second player would be able to claim a box first. But a savvy first player could exploit the strategy, sacrificing a single box to win the rest.

will claim by drawing a single line (hence "double cross"). In exchange, you'll receive the whole region that your opponent was eyeing.

skip this move, leaving it for your opponent

Beyond this level of strategy, as you seek to control the sizes and structures of the regions formed, it all gets murky and complicated. For such details, consult the writings of the late, great Elwyn Berlekamp. He passed away while I was writing this book and shall always be remembered as a child of otherworldly sophistication.

WHERE IT COMES FROM

Today, you'll find Dots and Boxes played just about everywhere: on blackboards, whiteboards, cardboard, legal pads, restaurant napkins, and, under desperate circumstances, bared arms.[3] It was first published by the mathematician Édouard Lucas, in 1889, under the title *La Pipopipette*. Édouard credited its invention to several of his former students at Paris's prestigious École Polytechnique.

This raises questions. Why would serious students spend their time crafting a game fit for children? And why would an esteemed scholar like Édouard choose to publish it?

Simple: Because serious mathematics is often born from childish play.

We see this pattern in Édouard's own career. He is perhaps best known for his work on Fibonacci-like sequences, in which each number is the sum of the previous two. (The classic sequence begins 1, 1, 2, 3, 5, 8, and so on.) Fibonacci numbers seem like a silly game—that is, until you start counting the bumps on a pinecone, the petals on a daisy, or the fruitlets on a pineapple, and realize that this silly game is played not only by children (and by adults of dubious maturity) but by nature itself.

---

3   It is also played in many countries, under names ranging from the pedestrian (the US's Dots; England's Squares) to the melodious (France's Pipopipette; Mexico's Timbiriche) to the visionary (the Netherland's Kamertje Vehuren, or "rent out a small room"; Germany's Käsekästchen, or "little cheese boxes").

Or take the cannonball problem, another of Édouard's favorites. It asks for a special number of cannonballs: one that you can arrange into a perfect square, and then rearrange into a perfect square pyramid. The puzzle is utterly frivolous. It is also devilishly hard. Édouard conjectured that the known solution (4,900 cannonballs) was the *only* solution.

Decades later, advanced work with elliptic functions finally proved him right.

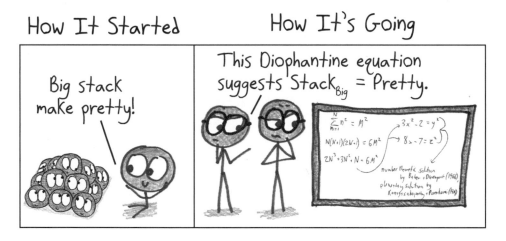

Or consider Édouard's most celebrated invention: the Tower of Hanoi. You may have seen one. It has three rods and a set of disks stacked from the largest on bottom to the smallest on top. Your goal is to transport the whole stack from one rod to another, moving one disk at a time, and never placing a larger disk on top of a smaller one.

In appearance and spirit, the tower is—how do I put this kindly—a baby toy. Yet it has found a variety of practical uses. Psychologists deploy it to test cognitive abilities; computer science professors, to teach recursive algorithms; and software engineers, as a rotation scheme for backing up data.

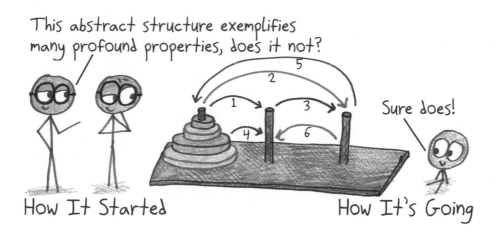

This abstract structure exemplifies many profound properties, does it not?

5
2
1
3
4
6

Sure does!

How It Started

How It's Going

How does recreation blur so easily into research? Why is the boundary between work and play so porous and heavily trafficked?

I honestly don't know. I don't think Édouard did, either. All we can say is, time and again, simple mathematical premises deliver profound consequences. That's what mathematics is, really: simple ideas in complex interplay. As Édouard wrote of Dots and Boxes: "Its practice, although easy, gives rise to continual surprises."

WHY IT MATTERS

Because useless play often births the most useful insights.

In his initial publication on Dots and Boxes, Édouard Lucas indulged in a long tangent about the value of pure curiosity. Citing a litany of historical examples, he argued that we must pursue questions for their own sake, no matter how silly they seem, because we never know what deep truths we might uncover.

His writing is flowery enough to be mistaken for perfume, but it's still worth quoting:[4]

When the ancients in dry weather rubbed a piece of amber with a cat's skin to attract light... they hardly suspected that this fact... which was for them a pastime... would be the germ of electricity theories, and many applications that astonish humanity...

---

4   By the way, if you ever feel like this book is beating around the bush, please consider that Édouard lays all of this out *before even getting to the game.*

When the geometers of Greece... cut across a very round and sharp root to study the shape and properties... they did not believe that their studies would serve more than twenty centuries later to help Kepler to formulate the laws of the motion of the planets...

When the priests of ancient Persia composed with the letters of the word Abracadabra... they had no idea that this symbolic painting would one day be taken up again by Tartaglia and Pascal in the form of the Arithmetic Triangle which is the foundation of modern algebra...

All mathematicians strive to uncover deep connections between disparate ideas. The question is how to do it. Hard work? Maybe. Patient calculation? Can't hurt. Looking up the answers in the back of the book? Sorry, you're getting colder. Exuberant leaps of imagination?

Now we're talking.

Édouard believed that profundity comes from play, science from silliness. He wasn't alone in this conviction. Elwyn Berlekamp learned Dots and Boxes at age six, and 70 years later, he was still playing it. The game lasted him a lifetime. Somewhere in the middle, while studying electrical engineering at MIT, it dawned on him that he could use mathematics to transform the game into an equivalent "dual game," which he called Strings and Coins.

How does this alternate version work? Picture a collection of coins, held together by bits of string. Each string has one end glued to a coin, and the other glued to another coin (or to the table). Players take turns cutting bits of string with the scissors. If your cut frees a coin, you pocket the coin, and cut again. When the last coin is freed, whoever has pocketed more coins is the winner.

No boxes, just coins. No drawing lines, just snipping strings. Yet the game is fundamentally the same. Without changing its core structure, Elwyn had turned Dots and Boxes inside out.

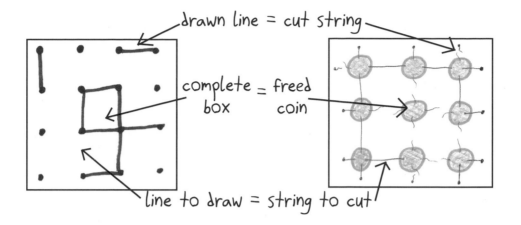

What's the point? No point. It's just cool. "Let the thinkers think and the dreamers dream," wrote Édouard Lucas, "without worrying whether the object of their attention seems sometimes useful, sometimes frivolous, because, as the wise Anaxagoras said, *everything is in everything.*"

That philosophy has driven millennia of mathematical inquiry, and it will last us for millennia to come. Let the thinkers think. Let the dreamers dream. Let the students doodle during lecture. Don't police imaginary borders between the practical and the impractical, the pointed and the pointless, the idle and the ideal. They all belong to the same vast continent, the same gorgeous wilderness that we have scarcely begun to explore.

VARIATIONS AND RELATED GAMES

SWEDISH BOARD: Begin with all lines along the outer rim of the board already drawn.

THE SWEDISH BOARD

THE SWEDISH, BORED

Dots and Triangles: All rules remain the same, except you play on a pyramid of dots, vying for possession of little equilateral triangles. In my view, this freshens up the game beautifully (and the pyramids aren't too tough to draw). Perfect for when you're growing bored with the classic version and your meal still hasn't arrived.

Nazareno: In this clever variant from Andrea Angiolino's *Super Sharp Pencil and Paper Games*, all rules remain the same, except two. First, on each turn, you can **draw a straight line of any length that you wish**, as long as it does not retrace any existing lines. (Thus, you may complete and claim multiple boxes with a single line.) And second, there is no bonus turn for completing a box.

Whereas Dots and Triangles' new appearance disguises the same basic game, Nazareno is the opposite: a familiar appearance disguising a fundamentally different experience.

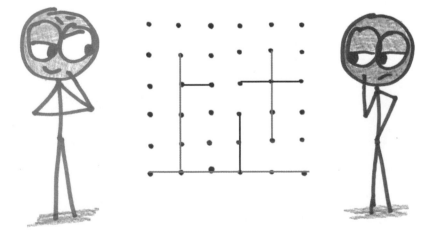

Square Polyp: In his book *100 Strategic Games for Pen and Paper*, the mad visionary Walter Joris offers several games reminiscent of Dots and Boxes. My favorite is 90: Square Polyp. To play, you need two players and two colors of pen.

1. Draw a 9-by-9 array of dots (or smaller for beginners/larger for experts), and **take turns placing square polyps**. These are squares with two adjacent lines poking out, like these:

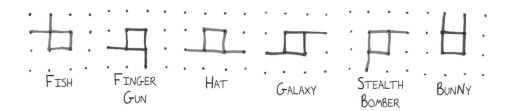

FISH   FINGER GUN   HAT   GALAXY   STEALTH BOMBER   BUNNY

2. **Claim territory by enclosing it entirely in your color.** Each polyp will automatically claim a 1-by-1 square, but with clever play, you can claim larger, more oddly shaped regions.

3. **Overlapping lines are forbidden.**[5] This makes it possible to foil your opponent's designs with a single stabbing tentacle (and for them to foil yours with the same ease).

4. **Play until no legal moves remain.** Whoever encloses a larger total territory wins.

winner!
20

13

---

5   If you want a variant on a variant, play tester Valkhiya suggested a neat twist on this rule: When placing polyps, you can overlap your own lines, but not your opponent's.

SPROUTS

A GAME OF "CURIOUS TOPOLOGICAL FLAVOR"

School geometry teaches us an ugly lesson: Size matters. In fact, size is the essence of matter. Angles can be acute, right, or obtuse. Figures can have length, area, or volume. Salted caramel mochas can be tall, grande, or venti. All of these traits boil down to size. Heck, the very name of the subject—"geo" meaning earth and "metry" meaning measurement—is about sizing up the world itself.

Does this size-conscious philosophy offend you? If so, you'll like topology. Its shapes stretch like rubber, squish like Play-Doh, and puff up like balloons. They're not shapes, really, but shapeshifters. In this oozing, lava-lamp world, size doesn't matter. In fact, "size" doesn't even mean anything.

Topology seeks deeper truths.

There's no better introduction to these truths than a game of Sprouts. Which spots can be connected? How many regions will form? What's the difference between "inside" and "outside"? Hold on to your hat—or the topological equivalent thereof—and enjoy a game that any child can play, yet no supercomputer can solve.

HOW TO PLAY

What do you need? Two (or more) players, a pen, and paper. Start by drawing a few spots on the page. For your first few games, three or four spots are plenty.

What's the goal? Make the final move, leaving your opponent with no viable options.

What are the rules?

1. On each turn, **connect two spots (or connect a spot to itself)** with a smooth line, and **place a new spot** somewhere along the line you just drew.

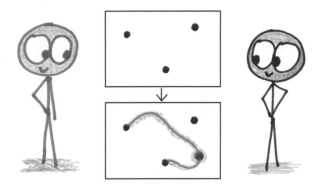

2. Just two restrictions: (1) **Lines cannot cross** themselves or each other, and (2) **each spot can have at most three lines sprouting from it.**

3. Eventually, you'll run out of moves. The **winner is whoever moves last.**

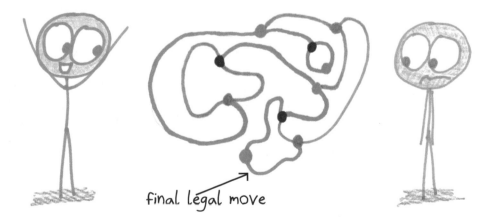

final legal move

TASTING NOTES

The delight of Sprouts is its flexibility. It doesn't matter whether you draw short segments, lazy curves, or mazelike spirals; all that matters is which spots you connect. You might even sign your name. A sixth grader named Angela busted out this move in our play-testing, and though it technically violates the "no crossing" rule, it was too awesome to disallow.

Angela's Handiwork

This flexibility captures the spirit of topology: Things that look quite different might be, for functional purposes, the same.

Consider a one-spot game. The first player must connect the spot to itself. After that, the second player must connect the two spots. It seems there are two different ways: through the inside, or around the outside.

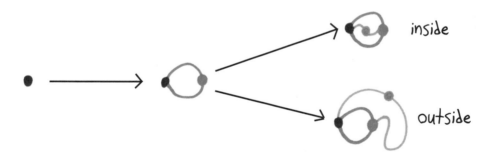

But wait. Imagine playing the game on the surface of a sphere. In this setting, nothing has changed, yet "inside" and "outside" have become arbitrary distinctions. The two moves are, for topological purposes, identical. The second player has no real freedom after all.

What about a game with two spots? Topologically, the opening move allows for just two choices: Connect the two spots, or connect one to itself. Whether you leave the other spot "outside" or "inside" is immaterial. In topology, it's all the same.

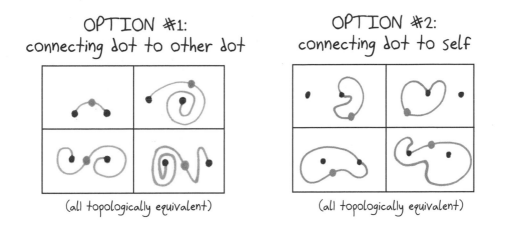

OPTION #1:
connecting dot to other dot

(all topologically equivalent)

OPTION #2:
connecting dot to self

(all topologically equivalent)

So do topologists ignore all distinctions, treating all things as identical? Is "winning" the topological equivalent of "losing"? Is "good" just topology-speak for "bad"? Is a cat topologically equivalent to a fish, and if so, should we place tiny litter boxes in our aquariums?

Well, that's up to you as a pet owner. But when it comes to Sprouts, you needn't worry. Not all moves are alike. In fact, by the second move of the two-spot game, you already face six topologically distinct options. The freedom grows from there.

ways the first two moves can unfold

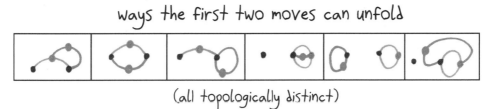

(all topologically distinct)

Dots and Boxes gave us rigid, rectilinear geometry, like a city laid out on a grid. Sprouts, by contrast, is an open-ended and free-form game, like the chaos of a traveling carnival.

WHERE IT COMES FROM

We can pinpoint the precise place and time of Sprouts' birth: Cambridge, England, on the afternoon of Tuesday, February 21, 1967.

Its parents, computer scientist Mike Paterson and mathematician John Conway, were doodling on paper, trying to invent a new game. Mike proposed the "add a new dot" rule, John proposed the name, and Sprouts was born.[6] The happy parents agreed to split the credit 60/40 in Mike's favor, a division

---

6 For whatever reason, pondering this game's name turns otherwise sober-minded people into baffling loons. One graduate student, noting that it involves spots and is contagious, suggested calling it Measles. Later, the otherwise insightful Eric Solomon wrote that the name Sprouts derives from the completed game's resemblance to "an over-cooked and disintegrating sprout," which is weird on two levels: First, that's not where the name comes from, and second, Eric Solomon should really consider letting someone else cook the sprouts.

so amicable and precise that it's almost more impressive than creating the game in the first place.

Sprouts is simple to play, yet almost impossible to solve. Mastering the six-spot game required a 47-page analysis from Denis Mollison. That was the state of the art until 1990, when a computer at Bell Labs solved games up to 11 dots. As of this writing, the largest solved game is over 40 dots—although Conway, before his death in 2020, questioned the legitimacy of this claim. "If someone says they've invented a machine that can write a play worthy of Shakespeare, would you believe them?" he asked. "It's just too complicated."

Has this dizzying complexity scared off casual gamers? Not at all.

"The day after sprouts sprouted," Conway wrote, "it seemed that everyone was playing it. At coffee or tea times there were little groups of people peering over ridiculous to fantastic sprout positions . . . The secretarial staff was not immune . . . One found the remains of sprout games in the most unlikely places . . . Even my three- and four-year-old daughters play it," he added, "though I can usually beat them."

WHY IT MATTERS

Because, of all the branches of modern mathematics, topology is among the most (1) dynamic, (2) bizarre, (3) useful, and (4) beautiful.

That's a lot of adjectives, so let's rumble through them, one by one.

TOPOLOGY IS DYNAMIC. Topologists navigate a shapeshifting world of stretching fabric, molten metal, and swirling soft-serve ice cream. Everywhere they go, they seek *invariants*: traits and properties that somehow, through all the change and upheaval, remain the same.

The most famous invariant is the *Euler characteristic*. In the context of Sprouts, it boils down to a simple equation (this version courtesy of Eric Solomon): spots + regions = lines + parts.

This equation holds true for every possible Sprouts scenario, from the beginning of the game to the end, from the simplest to the most complex, whether you begin on two spots or 2 million. No matter what, the number of *spots* plus the number of *closed regions* will always equal the number of *lines connecting spots* plus the number of *separate parts*.[7]

This is typical of topology: Beneath wild flux, we find powerful regularities.

---

7   A "part" is any group of connected dots; it may be as small as a single dot.

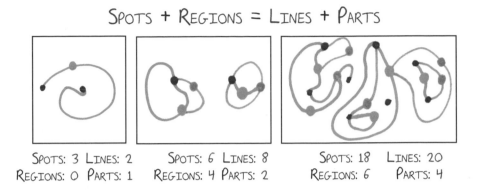

SPOTS + REGIONS = LINES + PARTS

SPOTS: 3 LINES: 2
REGIONS: 0 PARTS: 1

SPOTS: 6 LINES: 8
REGIONS: 4 PARTS: 2

SPOTS: 18 LINES: 20
REGIONS: 6 PARTS: 4

TOPOLOGY IS BIZARRE. Here's a fun result from John Conway. If a game of Sprouts lasts the minimum number of moves, then it must always end (roughly speaking) in one of these shapes:

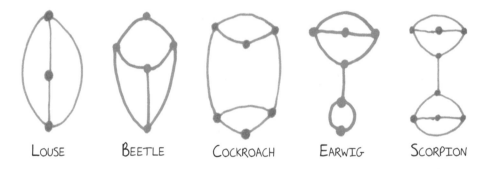

LOUSE   BEETLE   COCKROACH   EARWIG   SCORPION

As explained in the classic text *Winning Ways for Your Mathematical Plays*, the final configuration will consist "of just one of these insects (which might perhaps be turned inside out in some way) infected by an arbitrarily large number of lice (some of which might infect others)."

That's a lot of louses. Some configurations, as Conway quipped, are "lousier" than others.

TOPOLOGY IS USEFUL. Belying the silliness of lice and earwigs, topology yields insights about all sorts of things, from knotted DNA to tangled social networks, not to mention cosmology and quantum field theory.

Take a famous problem in topology: *graph isomorphism*. As we've seen, two positions in Sprouts may look different yet embody the same structure. How can we tell whether two networks are truly different or secretly the same?

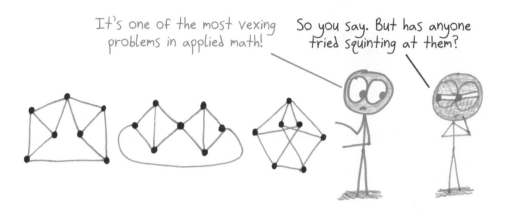

This question arises for electrical engineers comparing circuit schematics, for computer scientists encoding visual information, and for chemists looking up a compound in a database of structures. All of these sober scientists are in fact playing their own customized versions of Sprouts.

TOPOLOGY IS BEAUTIFUL. Many people meet topology through the surface known as a Möbius loop. Take a strip of paper, twist it halfway, and glue the ends together.

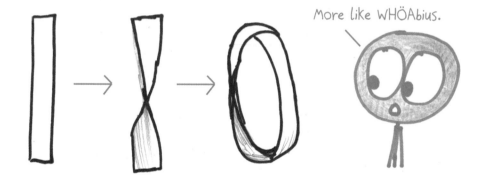

The Möbius loop has no "inside" or "outside." If you try to paint it like a bracelet, with a blue part facing your wrist and a red part facing the world, you'll fail. Whichever color you choose first will wind up covering the entire surface. And that's just one of its peculiarities. What happens if you cut a Möbius strip down the center? What if you cut it into thirds?

In his book *Euler's Gem*, mathematician David Richeson tallies up how many Fields Medals (math's most famous prize) have gone to topologists. "Of the forty-eight recipients," he writes, "roughly a third were cited for their work in topology, and even more made contributions in closely related areas."

If beauty comes from the marriage of simplicity and complexity, then Sprouts must be the favorite child.

VARIATIONS AND RELATED GAMES

WEEDS: Proposed by Vladimir Ygnetovich. On each turn, instead of adding a dot to the line you just drew, you may choose whether to add zero, one, or two dots.

POINT SET: In this variant from Walter Joris, play proceeds as in Sprouts, except that you score points by claiming regions. If your move creates a closed region, mark it with your initial or color, and score 1 point for each spot on the boundary of the region. No further moves in the interior of the region are allowed. When no legal moves remain, the winner is whoever has the most points.[8]

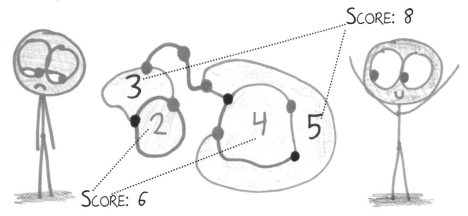

BRUSSELS SPROUTS: This evil twin of Sprouts seems, on the surface, to be every bit as open-ended and strategic as the original. It's not. In fact, it's less a game than a devious prank.

Begin with a few crosses, each with four free ends. Take turns connecting any two free ends, then putting a tick mark on the line you just drew, to generate two new free ends. Never cross an existing line. Last player to make a legal move wins.

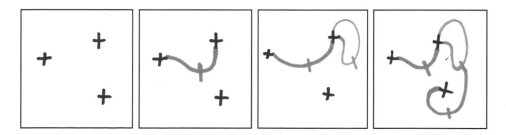

What's the prank? It's that your play has no influence on the game's outcome. With an odd number of crosses, the first player wins; with an even number, the second player wins. All your strategizing and scheming is no different than spinning a toy steering wheel and imagining that you control the car.

---

8   One extra rule is necessary: You are forbidden to create a closed region that contains a separate free-floating part within it, even if that part is just a single dot.

How can this be true? Well, note that the number of free ends never changes. Each move uses up two, then replaces them with two more. Instead, all that changes is the number of regions. Every move creates a new region—except for a few special moves. On a game with n crosses, no region is created by the $n - 1$ moves that connect previously unconnected crosses.

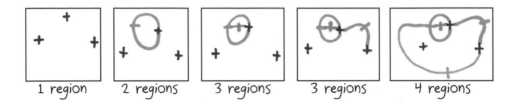

| 1 region | 2 regions | 3 regions | 3 regions | 4 regions |

The game ends when the number of regions catches up with the number of free ends. That requires $4n - 1$ region-increasing moves, plus the $n - 1$ non-region-increasing moves, for a total of $5n - 2$ moves.

CROSSES	+	+ +	+ + +	+ + + +	+ + + + +
MOVES	3	8	13	18	23
WINNER	First Player	Second Player	First Player	Second Player	First Player

To punk a friend, challenge them to games on two, four, and six crosses, each time generously insisting that they go first. When they smell a rat and demand that *you* go first, surreptitiously switch to a three- or five-cross game. Or, you know, don't cheat your friends! Either way.

ULTIMATE TIC-TAC-TOE

A GAME OF FRACTAL STRUCTURE

In 2013, after coming across this game at a math department picnic, I wrote a quick blog post about it. That post launched a brief internet phenomenon, reaching the top of Hacker News,[9] hitting the front page of Reddit,[10] and spawning a mini-industry of phone apps.[11] Since I owe my career in no small part to this game, I've thought a lot about what makes it special. Is it the elegance of the rules? The ease of developing strategic ideas? The subconscious association with Ultimate Frisbee?

Over the years, I've come to credit something else, something I should have suspected all along: fractals.

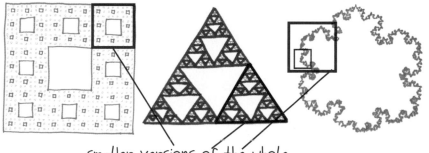

smaller versions of the whole

From clouds to coastlines to tree branches, we live surrounded by fractals. Perhaps that's why Ultimate Tic-Tac-Toe feels so natural. It's the game tic-tac-toe has always aspired to be.

THE RULES

What do you need? Two players, pen, and paper. Draw a large tic-tac-toe board, and then fill each of the nine squares with a smaller tic-tac-toe board.

What's the goal? Win three boards in a row.

How do you play?

---

9   If you don't know Hacker News: This is impressive.
10 If you don't know Reddit: This is also impressive.
11 If you don't know apps: This is not that impressive.

1. Take turns marking individual squares. The first move of the game can occur anywhere; after that, you must play on the mini-board dictated by your opponent's previous move. How so? **Whichever square they picked, you must play on the corresponding mini-board.**

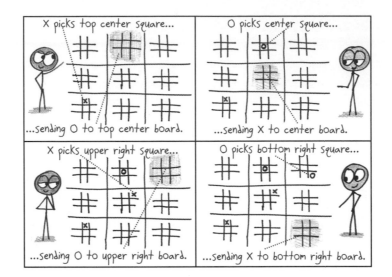

2. If you place **three in a row on a mini-board,** then you win that mini-board. The board is now closed; any player sent there may move anywhere they like, on any other board.

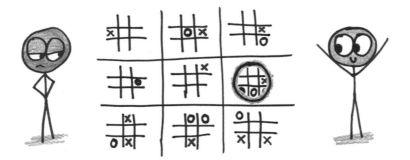

3. **Three mini-boards in a row wins the game.**

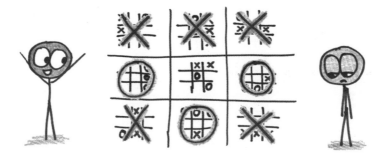

For other possible win conditions, see Ultimate Tic-Tac-Toe's VARIATIONS AND RELATED GAMES.

TASTING NOTES

One day in May 2018, I visited the political news site *FiveThirtyEight* to find a surprising headline. "Trump Isn't Playing 3D Chess," declared the lead story, written by Ollie Roeder. "He's Playing Ultimate Tic-Tac-Toe."

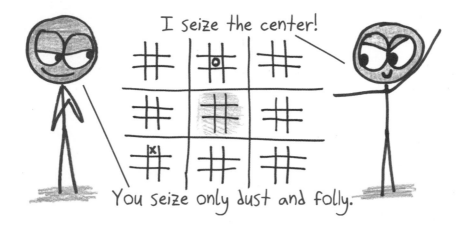

Back then, we all spent a lot of time analyzing the behavior of President Donald Trump. He'd leap from one political battle to the next, changing the subject at a whim. Was he carrying out a master plan? Or just obeying wild impulses? "He's not playing three-dimensional chess," critics would often quip.

Ollie Roeder agreed. In his view, Trump was playing another game entirely.

Whereas chess has just one battlefield, Ultimate Tic-Tac-Toe has many. "Those battlefields interact with each other in weird and complicated ways," Ollie wrote. "Even good ultimate tic-tac-toe play looks haphazard, facile, and even plain stupid at first glance." It's a game of "fluid, shifting goals," lending itself to "the strategy of distraction, delay, misdirection, procrastination, and improvisation." In other words: a Trumpian media strategy.

Good politics? Maybe not. Good game? Absolutely. More than that, it's a nifty conception of space: a fractal vision, where choices resonate between large and small levels.

That creates an inherent tension. What seems like a good move on the little board (such as taking the central square) may turn out to be a mistake in the grand scheme (by sending your opponent to the center board). To win, you've got to balance the two levels, doing what political activists strive to do: "Think globally, act locally."

WHERE IT COMES FROM

The earliest version I can find is a 1977 board game called Tic Tac Toe Times 10. A later version titled Tic Tac Ku won a 2009 Mensa Select award, with rules that differ slightly (you win not with three boards in a row, but by claiming five of the nine boards overall).[12] An electronic version titled Tic Tac Ten surfaced a few years later, with a rule change that speeds things up: If you win a single mini-board, you win the game.

Still, for whatever reason, my 2013 blog post marked the game's entry into the popular lexicon.

The game goes by many names. Wikipedia mentions "super tic-tac-toe," "strategic tic-tac-toe," "meta tic-tac-toe," "tic-tac-tic-tac-toe-toe," and "(tic-tac-toe)^2," omitting two others that I've heard: my favorite, "fractal tic-tac-toe," and my least favorite, "tic-tac-toe-ception."[13] In any case, the name "Ultimate" seems to have stuck. That's a point of immense pride for me, since it was coined by my students at Oakland Charter High School. Go Matadors!

WHY IT MATTERS

Because we live in a fractal world.

A fractal is something that looks the same on different scales. It is indifferent to zooming in, impervious to zooming out. See how tree branches split into smaller branches, each a miniature version of the whole? Or how coastlines trace jagged curves, appearing the same at all different scales? Even the fluffy architecture of clouds has a fractal quality.

It's no accident that we find these things beautiful. A simple design principle, repeated across scales, creates an enchanting complexity. It's what James Gleick, author of *Chaos*, calls "a wavering, lurching, animating harmony."

---

12 Their rules also state that if you are sent to a board that has already been won, you must play there, even though you can no longer affect the outcome on that board. This seemingly innocuous change turns out to break the game, for reasons I discussed in my first book *Math with Bad Drawings*. For best results, you should treat an already-won board as closed.

13 I believe that every linguistic pedant is allowed to wage one doomed, quixotic battle. Only one. If you rage about the nonliteral "literally," you can't also throw punches over "irregardless." If you're willing to die on the hill of "data" being plural, then you can't simultaneously die on the neighboring hill of "begs the question" vs. "raises the question." You must pick the fight that matters most to you, the one on which you believe civilization depends. My fight is *inception*. In the noisy, impressive 2010 film of that name, "inception" referred to planting a thought in someone else's mind, making them believe the idea was their own. This is a spectacularly useful word. It is what I seek to do to everyone in my life, just as they simultaneously (and more successfully) do it to me. Alas, the film's memorable climax involved a nested dream-within-a-dream-within-a-dream structure. Thus, people began using "inception" to mean "a thing within a thing within a thing." Pizza topped with mini-pizzas became "pizza-ception." In my view, this use of "inception" is doubly stupid, because it crowds out the proper name for the nesting concept ("fractal pizza") while leaving the thought-planting concept nameless ("pizza-ception" should mean "planting thoughts of pizza"). Anyway, if you agree, please photograph this footnote and tweet it at your legislators. I don't think they can do anything about it, but I like inundating them with weird tweets.

"A river is, in its essence, a thing that branches... its structure echoing itself on all scales, from river to stream to brook to creek to rivulet, branches too small to name and too many to count."

—James Gleick

Fractals arrived at the mathematical party in the 19th century, uninvited and somewhat unwelcome. These new shapes were jagged, fractured, and hard to visualize. Mathematicians called them "pathological," because they broke every rule of polite geometry.

For decades, though, no one grouped them together. They were just a scattered population of misfit toys. Then, in the 20th century, the mathematician Benoit Mandelbrot united them under the name "fractal" and began treating them not as the disease, but the cure. The cure to what? Well, to the crazy old idea that triangles, squares, and pyramids have something to do with physical reality. The real pathology, according to Benoit, was the geometry we're taught in schools. "Clouds are not spheres," he wrote, "mountains are not cones, coastlines are not circles, and bark is not smooth, nor does lightning travel in a straight line."

Nature is not Euclidean. It's fractal.

EUCLIDEAN WORLD

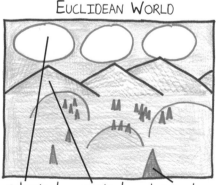
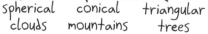

spherical clouds   conical mountains   triangular trees

ACTUAL WORLD

fractal clouds   fractal mountains   fractal trees

Plato would have hated it. The ancient philosopher believed so firmly in pure Euclidean geometry that one of his dialogues posits the whole universe is made of triangles—specifically, the two "special right triangles" that fill the nightmares of trigonometry students.

Well, okay, Plato. Go flip through your favorite nature account on Instagram. How many 30°-60°-90° triangles do you see?

Now look for fractals. A little more common, aren't they?

Nature is a garden of fractals. Mountains are jagged piles of rock, topped with smaller piles of rock, topped with even smaller piles of rock. Your lungs, beginning with the trachea, split and split and split again, an average of 23 times, before terminating in tiny balloon-like alveoli that feed oxygen into the blood. In short: You breathe fractally. Decades before fractal geometry was born, geologists realized that tiny streambeds and enormous canyons look indistinguishable in photographs, and so they always made sure to stick a lens cap or hammer in the frame for scale.

Every small thing is a microcosm, every large thing is a macrocosm, and every scale is an echo of every other scale.

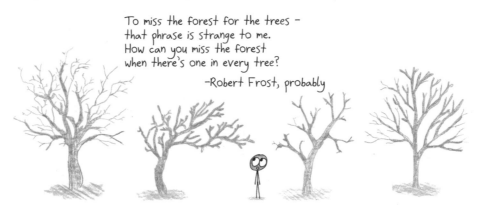

To miss the forest for the trees –
that phrase is strange to me.
How can you miss the forest
when there's one in every tree?

−Robert Frost, probably

To be precise, of course, the tree outside my office window doesn't branch infinitely many times. Maybe eight, tops. Still, according to *Fractal Worlds* by mathematician Michael Frame and poet Amelia Urry, that's enough. You earn the name "fractal" by having at least three self-similar layers. Ultimate Tic-Tac-Toe, being a square of squares of squares, qualifies. If you want to go one level deeper, combining nine such games into a 729-square board, be my guest.[14]

I'll admit that Ultimate Tic-Tac-Toe lacks the drama of forked lightning. It's an artificial fractal, just like the ones in human-made capacitors, or the plays of Tom Stoppard, or the paintings of Salvador Dalí. Still, like all works of human ingenuity, Ultimate Tic-Tac-Toe draws from the deep well of nature—a well that's brimming with fractals.

---

14 How would the rules work? Perhaps your move would be determined by the prior two: the penultimate move (which you made) determining which mid-sized board you play on, and the previous move (which your opponent made) determining which mini-board you play on. There's a world record waiting for anyone who cares (and dares) to try this.

"Fractal structures may occur from the Planck length to the size of the entire universe," Michael and Amelia write, "and maybe to the bubbling, branching growth of all universes. As far as we know, a larger range of scales is, literally, impossible."

Perhaps my students had this in mind when they gave fractal tic-tac-toe its most fitting moniker: "ultimate."

VARIATIONS AND RELATED GAMES

SINGLE VICTORY: First to win any mini-board wins the game.

MAJORITY RULES: To win the game, you must win more mini-boards than your opponent. Their arrangement does not matter, only their number.

SHARED TERRITORY: In the usual game, a mini-board that fills up without forming a three-in-a-row counts for neither player. But if you like, you may count it for *both* players (thereby making it easier to win three boards in a row).

ULTIMATE DROP THREE: Ben Isecke sent me this idea, a pleasing cross between Ultimate Tic-Tac-Toe and Connect Four. Play proceeds as in Ultimate Tic-Tac-Toe, except that wherever you place your X or O, it "drops" as far as possible in its mini-board. Making three in a row on any mini-board wins the game. (Alternatively, you could require victory on three boards in a row.)

On each turn, you have only three options: left, center, and right. The result is a tighter, tenser, more pressurized game, but still plenty complex.

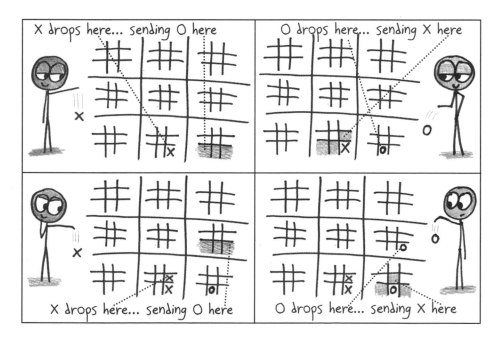

THE DUAL GAME: In the original game, your opponent's move determines the mini-board on which you must play. The dual game reverses this. Now, your opponent's move determines the *square* in which you must play; then, the choice of *mini-board* is up to you.

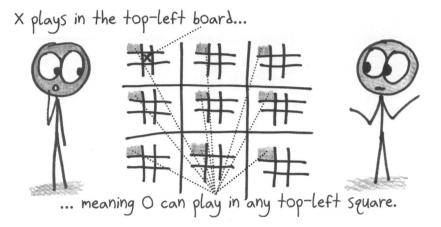

X plays in the top-left board...

... meaning O can play in any top-left square.

Here's one way to think of it. The original game sends you to a city, and you then choose a neighborhood. This version prescribes a neighborhood, and you can then occupy that neighborhood in any city you choose.

It's a tough game. I struggle to think more than a move ahead. Make sure to keep track of the most recent move; it's easy to get lost!

DANDELIONS

A GAME OF SPACE, TIME, AND OTHER SUCH FLUFF

I know your dreams, my friend. You wish to be a dandelion, riding the winds, a sentient piece of fluff borne across the fields of—

No, wait, I'm sorry. Misread that dream. You wish to be the wind itself, sweeping the fluff from the dandelions and carrying it—

No, wait. I see now. You want to be . . . both?

Aha! I have just the game for you.

HOW TO PLAY

What do you need? Paper, pen, and two players: the dandelions and the wind. To set up, draw a 5-by-5 "meadow" and a little compass rose.

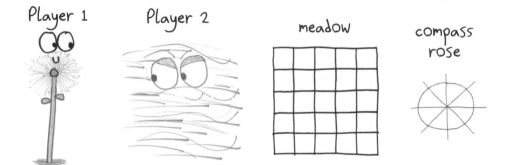

Player 1   Player 2   meadow   compass rose

What's the goal? The dandelions aim to **cover the whole meadow**. The wind aims to leave at least one square of the meadow uncovered.

What are the rules?

1. The dandelions move first, by **placing a flower** (i.e., an asterisk) anywhere on the grid.

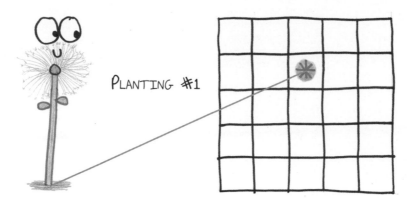

PLANTING #1

2.   The wind moves next, **choosing a direction in which to blow a gust that carries the dandelions' seeds**. Any vacant square downwind of a dandelion is now occupied by a seed (i.e., a dot). During the game, the wind may blow **only once in each direction**, so after a direction is used, mark it off on the compass rose.

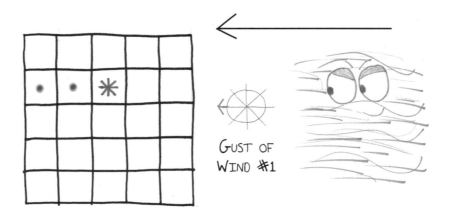

GUST OF
WIND #1

3.   Continue taking turns. **A dandelion is planted** (either in an empty square, or on top of an existing seed) . . .

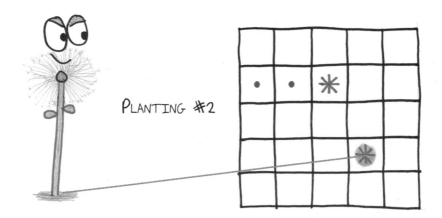

PLANTING #2

. . . and then the wind blows in a new direction, **carrying the seeds of all dandelions on the board,** and planting them in downwind squares. Note that **seeds emerge from all dandelions present,** but not from other seeds.

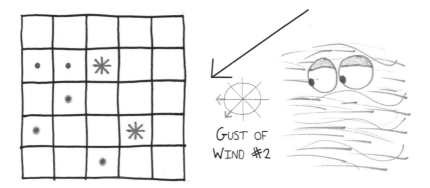

4. **The game ends after seven turns,** when the wind has blown in every direction except one. If the dandelions and their seeds **cover the whole board, then the dandelions win.**

DANDELIONS WIN!

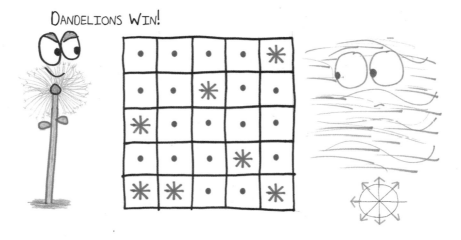

If any **blank squares remain, then the wind wins.**

THE WIND TRIUMPHS!

TASTING NOTES

"This game is as joyful as a field of yellow dandelions," writes Emily Dennett, one of my generous play testers, "and hopefully will spread just as fast."

The game belongs to a proud and sprawling family of asymmetric games, in which the players possess fundamentally different powers.[15] In the best of these, your sense of which side holds the advantage will change over time. In Dandelions, for example, many beginners find it easier to win as the wind, whereas experts tend to take the opposite view.

To this genre, Dandelions brings an extra twist: One player cannot avoid helping their opponent. They can only try to be minimally helpful. As play tester Jessie Oehrlein put it: "The player that wants to spread can't spread, and the player that wants to block can't block."

WHERE IT COMES FROM

I was working on a game called Paint Bomb when my pal Ben Dickman challenged me to evolve beyond the warlike theme. My first thought was to retitle it Dandelions. That didn't fit the old game (now called Splatter), but it lent itself to a whole new one: a game of windswept seeds and open meadows, a game of simultaneous competition and collaboration. This game.

WHY IT MATTERS

Because every spatial game is a temporal game, too.

In my early experiments with Dandelions, the wind always won. It seemed impossible to fill every square. Then I realized something: Once two dandelions are in play, certain squares become guaranteed. For example, if the wind has not yet blown south or east, then a square to the south of one dandelion and the east of another will, sooner or later, gain a seed. The wind can't prevent it.

---

15 In many—such as Fox and Hounds, the Nepali classic Bagh-Chal, and the Tafl games from Scandinavia—one player seeks to escape, while the other seeks to capture.

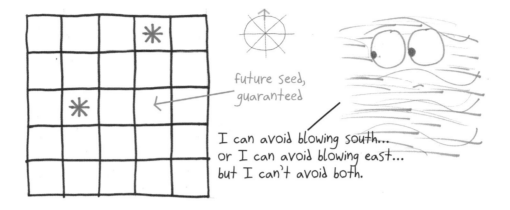

future seed, guaranteed

I can avoid blowing south...
or I can avoid blowing east...
but I can't avoid both.

Slowly, I trained myself to identify those guaranteed squares, and then—an even harder step—to ignore them. Better to focus on the squares still in doubt. This new perspective demanded patience and, beyond that, an altered conception of time. I had to collapse the distinction between "already filled in the past" and "guaranteed to be filled in the future."

This changed my view of the game. It revealed, for example, that the first dandelion will benefit from seven gusts of wind, and the last dandelion, from just one. The early plantings thus exert a far greater influence on the outcome.

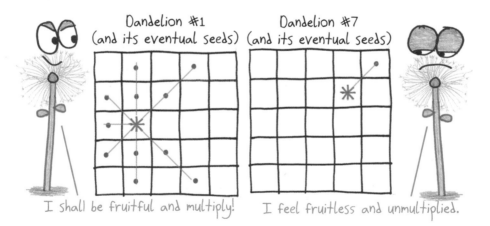

Dandelion #1 (and its eventual seeds)   Dandelion #7 (and its eventual seeds)

I shall be fruitful and multiply!   I feel fruitless and unmultiplied.

As for the wind, it's just the opposite. Because gusts tend to result in fewer new seeds as the board fills up, it's natural to view the early gusts as more impactful. But that's backward. The first gust carries the seeds from a single dandelion. The final gust carries seeds from seven. Later gusts matter more.

Gust #1
(and its resulting seeds)

Gust #7
(and its resulting seeds)

Though the game's structure is spatial, its subtleties are temporal. It shares that trait with chess and go, which for all their geometric intricacy are more often likened to conversations unfolding in time. Each is a back-and-forth exchange of ideas. The same is arguably true of geometry itself; there is no geometry without thought, and no thought without time.

Dandelions is "a quick game of spatial reasoning," wrote play tester Jonathan Brinley, "where you try to anticipate the long-range effects of a series of immutable decisions." To pull this off requires a willingness to turn time inside out.

Consider the collaborative variant, wherein the wind and dandelions work together to cover as much territory as possible. How best to conceptualize this task? By rearranging time altogether.

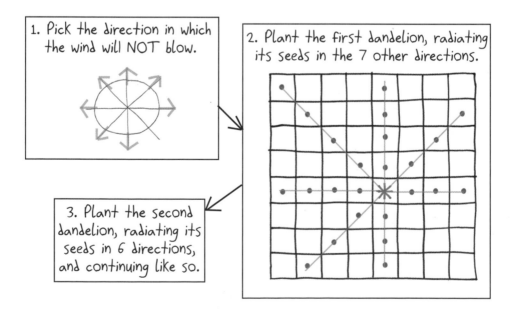

1. Pick the direction in which the wind will NOT blow.

2. Plant the first dandelion, radiating its seeds in the 7 other directions.

3. Plant the second dandelion, radiating its seeds in 6 directions, and continuing like so.

We tend to imagine space and time in fixed, absolute terms. Space is a box containing the universe's playthings. Time is a ticking clock on the wall.

But games force us to imagine other relationships, other marriages between the spatial and the temporal. In Dandelions, the past and future interlace.

Time and space play off of each other, just as the wind and dandelions do: asymmetric partners filling the meadow with their bright yellow offspring.

VARIATIONS AND RELATED GAMES

BALANCE ADJUSTMENTS: Asymmetric games often seem to favor one side over the other. To improve the wind's chances, play on a **larger meadow** (i.e., a 6-by-6 grid). To improve the dandelions' chances, let them begin the game with a **double planting** (i.e., two dandelions are planted), while the wind ends the game with a **double gust** (i.e., after the seventh planting, the wind blows twice).

KEEPING SCORE: Switch to a larger board (I suggest 7 by 7), so that the dandelions will struggle to cover it fully, and then take turns playing each role. The wind scores a point for each square left uncovered. Whoever scores more points overall wins.

RANDOM PLANTINGS: This solo variant was suggested by Joe Kisenwether. Play on a 6-by-6 board, rolling two dice (interpreted as x- and y-coordinates) to determine the placement of each dandelion. Then, you play as the wind, trying to cover as much of the meadow as possible.

RIVAL DANDELIONS: This idea comes from Andy Juell. Play on a larger meadow (at least 8 by 8), with each player planting dandelions in their own color. On each turn, **one player plants, then the other**. Alternate who chooses first. Once a square is filled by a seed or flower, it cannot be filled again.

After both players have planted, the **wind blows in a randomly chosen direction**, determined with an eight-sided die (easily simulated; do an internet search for "roll dice").

Whoever covers **more squares in their color** is the winner.

COLLABORATIVE: Using a larger meadow (such as 8 by 8), wind and dandelions work together to cover the whole board. To increase the challenge, Guillaume Douville suggests that you refuse to discuss strategy, communicating only through your moves, perhaps even playing in silence.

On sufficiently large boards, you may want to allow eight plantings and eight gusts of wind.

The cooperative game also works as a solo puzzle. What's the largest board that can be filled this way? On grids too large to fill, what's the maximum number of squares you can cover?

QUANTUM TIC-TAC-TOE

A CONFUSING GAME FOR A CONFUSING UNIVERSE

"Those who are not shocked when they first come across quantum theory," said the physicist Niels Bohr, "cannot possibly have understood it." Let these words serve as a warning: Quantum Tic-Tac-Toe is the trickiest game in this book. You'll need patience to absorb the idea of placing your X (or O) in two tentative squares. You'll need additional patience to master the "collapse" process, whereby your X winds up in one square or the other. And you'll need superhuman patience to brave intimidating vocabulary terms such as "entanglement," "superposition," and, most esoteric of all, "state."

Trust me: It's worth it. Awaiting you are strategic twists, surprising nuances, and, best of all, a spark of insight into the quantum realm itself.

HOW TO PLAY

What do you need? Two players, a pen, and plenty of paper.

What's the goal? It's just like classical tic-tac-toe: Place your entangled particles so that when the waveform collapses, you're left with three in a row.

Okay, maybe not *just* like classical tic-tac-toe . . .

What are the rules?

1. Take turns placing quantum X's and O's. To do this, **mark any pair of boxes, connecting them by a thin line**. The boxes (which are now "entangled") need not be adjacent. Later, your particle will end up in one of the two. Which one? For now, it's a mystery.

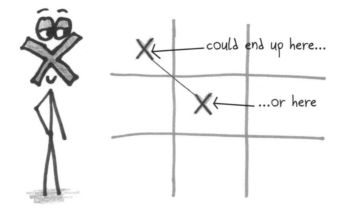

As you play, multiple quantum particles may seem to share the same box. But that's only temporary. **Eventually, every box will contain just one "classical" X or O**.

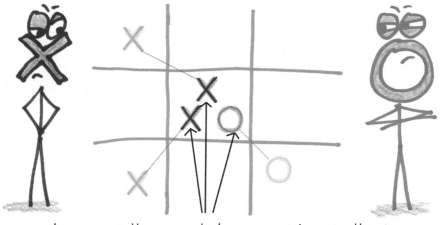

only one of these particles can wind up in this box

2. At some point, the **entanglements will form a loop**: For example, one box is entangled with another, which is entangled with another, which is entangled with the first.

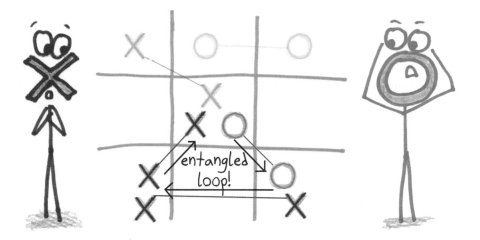

entangled loop!

At that moment, the particles **collapse**, assuming their fixed and final locations. This can **unfold in two ways**: one for each possible location of the last particle placed. Whichever way it goes, it will **force another particle out of its box, and into a different box**. This forcing process continues until every particle in the loop resolves into a single box.

If it goes here...

... that forces this particle here...

... which forces this particle to go here.

But if it goes here...

... that forces this particle here...

... which forces this particle to go here.

Some other particles may have just one "toe" in the loop. These will be forced out, into their other box.

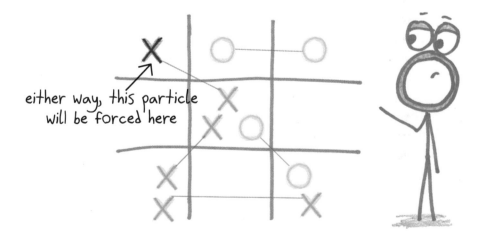

either way, this particle will be forced here

3. Someone must choose between the two ways that the collapse can unfold. **This choice falls to the person who *didn't* complete the loop.**[16] When the collapse is over, your board will be a total mess, so **redraw it** and continue playing.

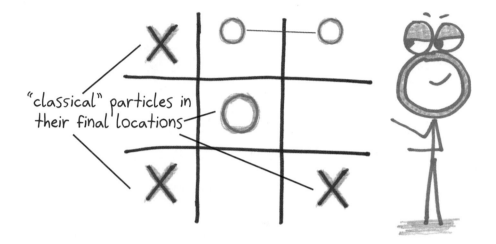

"classical" particles in their final locations

4. Note that "classical" particles are final; no new particles can be placed in those boxes. If you achieve **three classical particles in a row**, then you win!

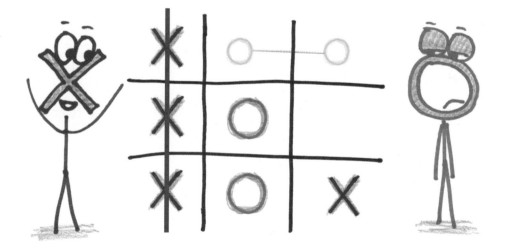

Two players may achieve three in a row as part of the same collapse. If so, then they **both win**. It's not a tie. It's a shared victory. Such is the strangeness of quantum life!

---

16 For a variant with more randomness, you can instead decide the direction by flipping a coin.

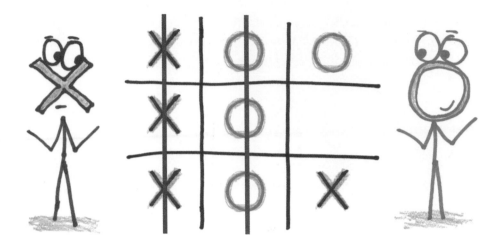

Finally, if eight of the nine squares contain classical particles, yet no one has won, then call it a draw.

TASTING NOTES

One play tester praised the rules for their "alien elegance." That captures the yin and yang of the game. To some, it's elegant; to others, just alien. Anyway, if you persist through the fog, strategic possibilities begin to emerge.

One sneaky tactic is to create short all-X (or all-O) cycles. Whereas completing a cycle is usually risky, because it lets your opponent control the collapse, you needn't worry if the whole cycle is made of your symbol. The quantum marks will simply become classical ones.

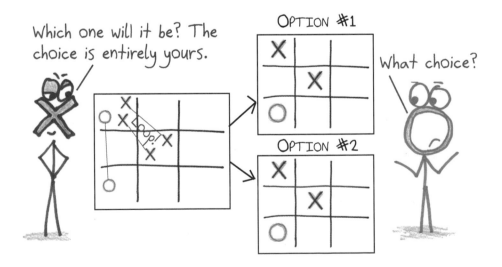

The threat of such a loop will often force your opponent to preempt you by completing the loop instead, giving you control of the collapse. That makes it a powerful stratagem. That said, it's more fun to create long strings of entanglement. You and your opponent might conspire to fill the whole board with a glorious nine-move loop, whose collapse (at O's discretion) brings the game to a dramatic finish.

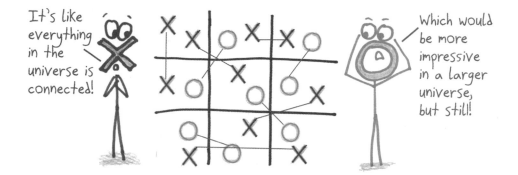

As in classical tic-tac-toe, X enjoys a strong advantage. I recommend playing a few games, switching roles each time, and keeping score as you go. Award double points if X manages to achieve a pair of three-in-a-rows on the same collapse.

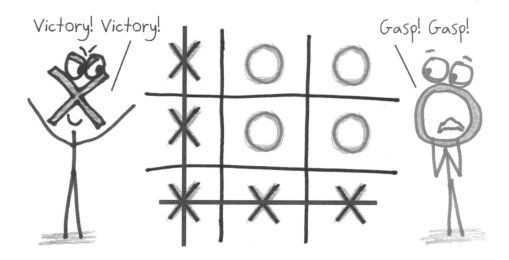

WHERE IT COMES FROM

The game was invented by software engineer Allan Goff. "After getting the idea," he wrote in a 2002 paper with two coauthors, "it took only 30 minutes to develop the rules. It felt more like a process of discovery than one of invention."

My favorite line in that paper: "Tic-Tac-Toe is a classic children's game that reinforces our prejudices toward a classical reality." I've heard a lot of charges levied against tic-tac-toe—too boring, too simple, too many draws—but it "reinforces our prejudices toward a classical reality" is a new one on me. It nicely illustrates the purpose of the quantum game: as a handy teaching tool for the counterintuitive concepts of quantum physics.

WHY IT MATTERS

Because the universe says so.

I hear your objections. Heck, I share them. This quantum game disturbs our conception of space, challenges our notion of time, and calls into question the very meanings of "tic," "tac," and "toe." In my darker moments, I suspect it is no game at all, but a headache with victory conditions. I wish the universe weren't like this.

But the cosmos has been quite clear. It wants us to play by quantum rules.

If the cosmos doesn't *feel* very quantum, that's because we're not 0.00000001 meters tall. Above that scale, the classical description of reality—with its solid objects, ticking clocks, and billiard-ball-like particles—works well enough. Below that scale, though, the rules break down, and new rules take hold.

For example, in quantum physics, particles don't have specific locations. The electron isn't here. The electron isn't there. The electron is *everywhere*, all at once, smeared across space, in a kind of probabilistic cloud. Whereas classical objects have *positions*, quantum objects have *superposition*, seeming to reside in multiple locations at once.

Will the rest of your party be arriving soon?

No. Or... yes? In a sense, they're already here.

I'LL HAVE THE RAVIOLI

THE DIFFICULTIES OF DATING AN ELECTRON

Why don't we ever see particles engaging in this duplicitous behavior? Because, weirdly enough, quantum behavior changes when observed. The moment you take a measurement, the particles react like misbehaving kids when the school principal arrives. The chaos comes to an abrupt end, the possibilities fall away, and each particle snaps into a single location.

Case in point: Schrödinger's cat. In this thought experiment, we imagine a cat in a box with a devious device. It detects whether one particular radioactive atom decays, and if so, it releases a poison, killing the cat. No decay? No poison, and the cat lives.

Now, before we open the box, the system remains unobserved. Thus, the atom is in a state of superposition: It has decayed, and it hasn't decayed, both at once, according to the probabilistic logic of quantum mechanics.

Therefore, the cat is dead, and the cat is alive, both at once—until we open the box, and the two possibilities collapse into one.

In Quantum Tic-Tac-Toe, "observation" occurs when you complete a loop. That's when the quantum weirdness ends and the squares collapse into regular tic-tac-toe moves. The unrealized possibilities vanish without a trace.

Or do they?

This question has troubled physicists and philosophers for a century. One viewpoint—the "many worlds interpretation" of quantum mechanics— holds that the so-called collapse never really happens. Instead of assuming one position or another, the particle assumes both, each in a different parallel universe. In one reality, the cat is dead. In another reality, it's alive. In one universe, your X lands in the corner. In another, it lands in the center. Existence is an explosion of parallel universes, branching countless times every nanosecond.[17]

---

17 You can even play Quantum Tic-Tac-Toe this way—see VARIATIONS AND RELATED GAMES for the details.

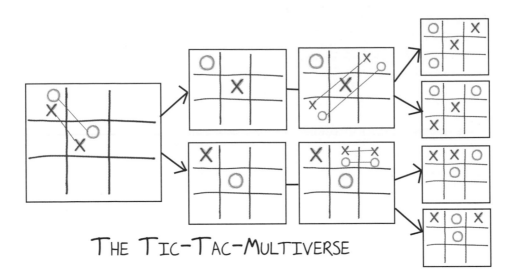

THE TIC-TAC-MULTIVERSE

Quantum Tic-Tac-Toe displays another mystifying feature of the quantum realm: *nonlocality*. When two squares become entangled, observing the outcome of one ("Oh, it's an X") gives you immediate knowledge of the other ("Hey, it must be an O"). Somehow, with no passage of time, and no physical contact, cause and effect emerge simultaneously, like instant messages sent between distant star systems.[18]

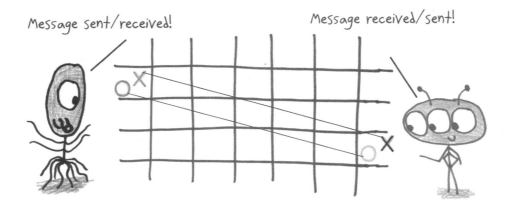

Message sent/received!   Message received/sent!

By the end of the game, the quantum weirdness is exhausted. The final board shows classical particles, traditional X's and O's. This constitutes our final lesson in quantum physics.

At large enough scales, the quantum ceases to look quantum at all.

You and your dog are technically composed of quarks and electrons. But you've got so many quarks, and so many electrons, that you aren't really quantum creatures. You don't occupy multiple locations at once. You don't

---

18 In fact, writer Ursula Le Guin elaborated this idea into the *ansible*, a fictional technology that allows for faster-than-light communication between planets.

change physical properties when observed. You don't even participate in faster-than-lightspeed causal loops.[19]

So it is with Quantum Tic-Tac-Toe. A giant game—played, say, on a 1,000-by-1,000 board—would undergo so many collapses by midgame that it would appear, from a distance, perfectly classical. Only by zooming in would you detect the strange quantum structure.

That's exactly how physical reality works. Classical at a distance, quantum up close.

We all know tic-tac-toe as a simple little game. Perhaps too simple. Perhaps too little. Yet that makes it just the right size for investigating the mysteries of quantum mechanics, mysteries that surround us, mysteries that live within us, mysteries that come into view only when we attend to the simplest and littlest things in existence.

VARIATIONS AND RELATED GAMES

MANY WORLDS: Ben Blumson suggested this variant. "When a loop is formed," he explained, "the board would be simplified as usual, but instead of one player determining which board to play, **the players would continue to play on BOTH boards**. The game would be won by the player who **wins in most branches**."

Once a particular board has been won, you may stop playing on it, and simply count it toward the winner. Since an early victory ought to count for more than a later victory, **make each board worth $(\frac{1}{2})^n$**, where n is the number of branchings it has undergone.

For a simpler version, you can require players to make **the same move on all boards**. This ensures that all boards will collapse simultaneously, and **each collapse will double the number of boards** (unless the loop is all X's or all O's, in which case the number of boards will remain the same).

TOURNAMENT STYLE: Play on a **4-by-4 grid**, and don't stop until the board is full. **Every three-in-a-row that you form scores 1 point.** A four-in-a-row, since it consists of two overlapping three-in-a-rows, scores 2 points. Most points wins. Played in CodeCup 2012 (and brought to my attention by Joe Kisenwether), this is a great next step once you have a handle on the game.

QUANTUM CHESS: Because a fully quantum chess would burst our brains like so many kernels of popcorn, this variant (brought to my attention by Franco Baseggio) applies quantum logic to only one piece: the king. Play proceeds as normal, until the first time you move your king. Then, instead of moving it, **place a coin on any square that the king might now**

---

19 Well, you don't, anyway. I can't speak for your dog.

occupy. It no longer exists in a definite location, but in a cloud of possible locations.

Later, if you move the king again, **place a coin on any square legally reachable from any of the king's previous positions**. If you wish to capture an opposing piece with your king, then you must commit your king to that specific location and remove all your coins from the board.

If your opponent places a king location in check, you may either (a) defend the location as you normally would, or (b) abandon the location by removing your coin. Such an abandonment does not count as a move; it just means the king was never there to begin with.

Checkmate occurs when the king has no safe locations remaining.

A CONSTELLATION OF SPATIAL GAMES

In the vast galaxy of spatial games, we have visited five planets. By my calculations, that leaves about 70 gazillion. Due to space constraints (see what I did there?), I see three options: (1) Investigate a single additional game in thrilling, tedious detail; (2) lay out five more games in breezy brevity; or (3) summarize 5,000 games in size 0.1 font.

Well, at the risk of coming across as an unimaginative and doughy-souled centrist, let's steer the safe middle path, shall we?

BUNCH OF GRAPES

A GAME OF HUNGRY FLIES

Whereas most pencil-and-paper games leave the paper covered in crisscrossing gibberish, this territorial struggle (designed by Walter Joris) ends up like a page from a coloring book, looking patterned, pretty, and good enough to eat.

To begin, draw a bunch of grapes. Make clear which grapes share a border. Then, one at a time, **each player marks a grape with a colored dot, to indicate where their "fly" begins**.

On each turn, **your fly consumes the grape it is on** (shown by fully coloring in the grape) then **moves to an adjacent grape**. Whoever placed the second dot gets to move first. In the end, whoever is **unable to move because there are no adjacent grapes available is the loser**.

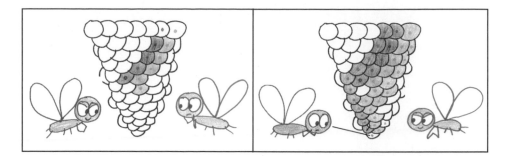

I expected this game to be dull and predictable, with the best move obvious at each step. Instead, it's full of surprises and close calls. I credit the grapes themselves: uneven in size, erratic in arrangement, and different every time, they deceive the eye, causing you to misestimate the space available. Bunch of Grapes is a spatial game in the truest sense, all about your perceptions (and misperceptions) of space. Best played while snacking on grapes.

NEUTRON

A GAME OF BACK AND FORTH

The titular "neutron" is a neutral particle, shunted back and forth between opposing teams, a kind of abstract hockey puck. But in this game, no skater knows how to stop, and the goal you want to score on is your own.[20]

You'll need a **5-by-5 grid** and **11 game pieces**: five of one kind, five of another, and one special marker (the neutron itself). The goal: **Get the neutron into your home row.**

On each turn, you first **move the neutron one step in any direction** (like a chess king),[21] and then move one of your pieces **as far as possible in any direction** (like a chess queen whose brakes have been cut and won't stop until it encounters an obstacle). The exception is the game's opening move: The first player doesn't move the neutron, just one of their own pieces.

---

20 Developed in 1978 by Robert A. Kraus, the game enjoyed a sudden resurgence in 2020 when a user on the site Board Game Arena uploaded it under the name Bobail. As far as I can tell, Bobail is just Neutron with one slight rule change—and seeing as Kraus presented 15 variants, I suspect this was one of his.
21 Actually, in Kraus's original rules, the neutron moves like the other pieces. This is the Bobail variation.

Opening Move   Second Move
                    (part 1)   (part 2)

You can win in one of two ways: (1) if the neutron reaches your home row, or (2) if you trap the neutron so that your opponent cannot move it on their turn.

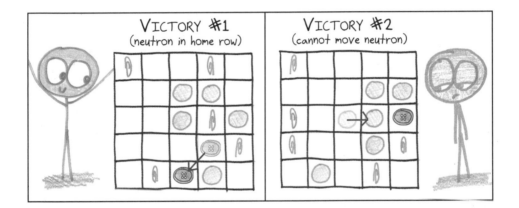

VICTORY #1
(neutron in home row)

VICTORY #2
(cannot move neutron)

You reach the game's depths sooner than you expect, as if wading into the water to find a plunging seafloor. I find it gratifying when I can force my opponent to move the neutron in my direction (or better yet, force them to push the neutron into my home row, thereby winning the game on their move). Meanwhile, it's hard to trap the neutron unless you've got the upper hand already; when on defense, you have fewer safe choices, so it's harder to spring a trap.

ORDER AND CHAOS

A GAME OF ELEMENTAL STRUGGLE

Published by Stephen Sniderman in a 1981 issue of *Games* magazine, this two-player game embodies an ancient conflict. It is the struggle of makers vs. breakers, structure vs. destruction, parents vs. children, Bert vs. Ernie.

It is the battle of Order and Chaos.

Play on a **6-by-6 grid**. One player (order) aims to create a five-in-a-row; the other (chaos) aims to prevent any such five-in-a-row. Players take turns marking squares, and **each is free to use either symbol (X or O) as they please**.

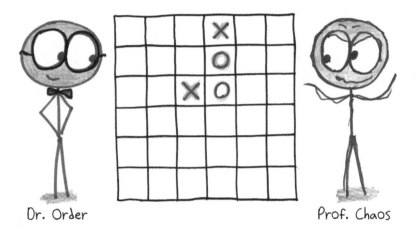

Dr. Order   Prof. Chaos

Order wins by achieving a five-in-a-row. The five can be horizontal, vertical, or diagonal, and made either of all X's or all O's.

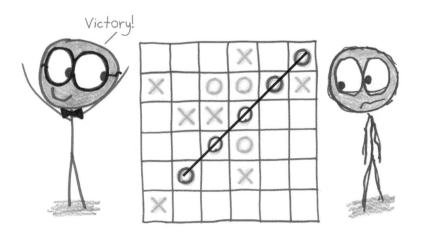

Chaos wins if the board fills up with no five-in-a-row created.

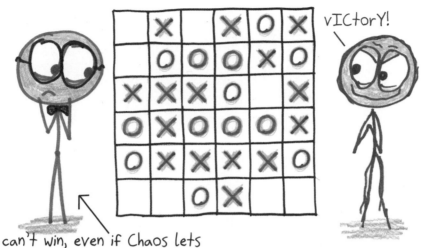

vICtorY!

can't win, even if Chaos lets
them fill the rest of the board

Deliciously, each player's symbols can be used against them. Order's old moves may block new progress, and chaos's may form part of an eventual five-in-a-row. The game is also nicely balanced: Beginners often feel that chaos has the advantage, while experts tend to consider order the stronger side.

I suggest playing several rounds, switching sides each time, and keeping score as you play. If you win as order, score 5 points, plus 1 point per blank square remaining. If you win as chaos, score 5 points, plus 1 point per blank square remaining *at the moment when the board becomes so clogged that order cannot possibly win.*[22]

Andy Juell suggests another fun variant: Once per game, order may place a special $\otimes$ symbol (which serves as both X and O), while chaos may place a ■ symbol (which serves as neither X nor O). I call these symbols "the jewels" in his honor. If you're finding the game imbalanced, you may level the playing field by giving the Juell Jewel only to the weaker side.

---

22 A way to identify this moment: If you filled all remaining spots with X's, would order win? What if you filled all remaining spots with O's? If the answer to both questions is no, then order is doomed. Stop the round and call it a victory for chaos.

SPLATTER

A GAME OF EXPLOSIVE PAINT

For this two-player game, you'll need a **rectangular grid of any size, filled with two kinds of paint blobs in equal numbers**. For a speedy setup, let one player fill the grid as they please; the other then gets to pick a color and go second (or defer the choice of color and go first). For a slower alternative, assign colors beforehand, and take turns placing blobs on an empty grid.

Now, on each turn, **splatter one of your paint blobs**. A blob can splatter in two ways: **alone**, or **taking all of its neighbors with it**. Either way, shade in the affected squares; they are eliminated from the game. Take turns splattering, never skipping a turn. **The last color with an unsplattered blob remaining is the winner.**

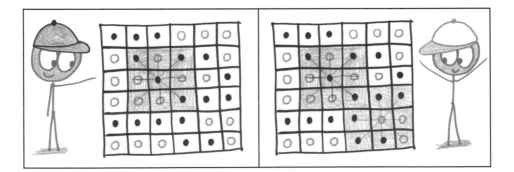

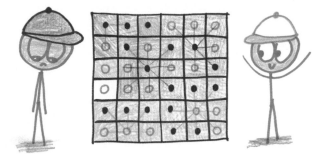

The game unfolds at its own peculiar tempo. Sometimes you want to accelerate the pace, splattering as many squares as possible. Later, you may want to apply the brakes, splattering lone squares in an effort to eke out extra turns.

For a more complex variant, allow two other splatter patterns: **diagonal** (splattering to the northwest, northeast, southwest, and southeast) or **orthogonal** (splattering to the north, south, east, and west). Each of these options will splatter four neighbors while leaving the other four untouched.

3D TIC-TAC-TOE

A GAME OF LENGTH, WIDTH, AND DEPTH

Unless you're reading this in a future filled with excellent VR technology—in which case, do you people still read books?—it takes a bit of a hack to picture the third dimension in this game. Instead of a cube, **draw four 4-by-4 squares**, one to represent each layer of the game.

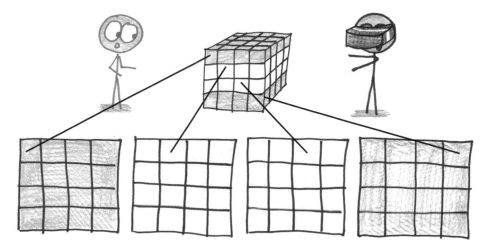

Take turns placing X's and O's. The first to create four in a row wins. Watch out for four-in-a-rows that slice across all the layers of the board. They're sometimes hard to spot until it's too late.

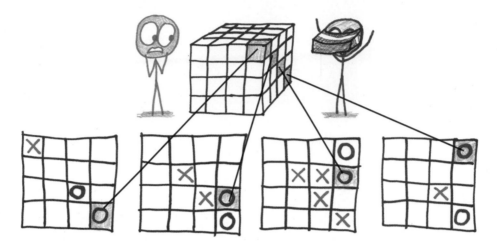

For strategic insight, you can measure the value of a square by counting the possible victories that pass through it. Here's what that tally looks like for standard tic-tac-toe:

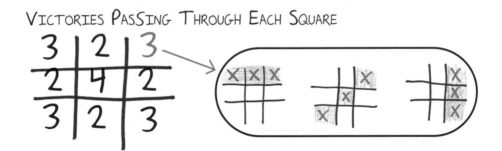

VICTORIES PASSING THROUGH EACH SQUARE

Applying the same method to the 3D game, an interesting pattern emerges. The best squares are the corners of the external boards and the centers of the central boards.

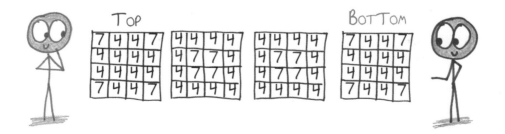

There's a fun meta-game here: What other classics can you expand into the third dimension? Some, such as 3D Battleship, barely need adjustment. Others, such as 3D Dots and Boxes, are easy to formulate (you claim a cube by drawing its twelfth edge) but hard to visualize (good luck drawing lines between layers). And some games, such as 3D Sprouts, stop working entirely (because in 3D, the lines no longer create separate regions, making the game as pointless and preordained as Brussels Sprouts).

Here's a suggestion to get you started: You can turn 3D Tic-Tac-Toe into 3D Connect Four, with pieces that "drop" to the bottom of the column, by simply requiring that every mark go either (a) in the bottom layer, or (b) one layer directly up from an existing mark.

Now you try!

II
NUMBER GAMES

BRACE YOURSELF, BECAUSE I'm about to hit you with an airtight philosophical proof that every number is interesting.

Every number. No exceptions.

Although I'd love to celebrate each number in turn—1 the loneliest, 2 the only even prime, 3 the best *Toy Story* film—that's a recipe for burnout. So let's stipulate that, somewhere along the line, we hit an uninteresting number.

After climbing from 12 (the number of distinct pentominoes) to 19 (the size of the only magic hexagon made from consecutive numbers) to 561 (the smallest absolute pseudoprime[1]), each number as unique as a child and as special as an ice cream sundae, we run smack into a dull number. We cube it. We factor it. We ask the band Three Dog Night to give it a yearbook superlative. Nothing works. This dullard is like no number we've seen. It is, for the first time, uninteresting. Isn't that surprising? Shocking, even? Might you even call it . . . oh, what's the word I want . . .

Interesting?

If there were uninteresting numbers, then there would be a first one. Yet the first uninteresting number would be very interesting indeed. Logic forbids such a paradox. Thus, all numbers must be interesting.

As the pros say: QED.

---

1   That is, for any number *n*, the power n^{561} will be precisely *n* more than a multiple of 561. The same can be said of all prime numbers, but 561 is the first such composite number. This will not be on the quiz.

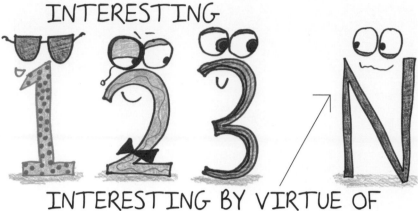

INTERESTING

INTERESTING BY VIRTUE OF
BEING UNUSUALLY UNINTERESTING

Wikipedia calls this proof "semi-humorous," a harsh burn by Wikipedia standards. Still, I believe it captures something about the spirit of math. I'm drawn to numbers for the same reason that millions of harried, overscheduled people set aside 15 minutes each morning for sudoku: not to put food on our tables, or Bitcoin in our pockets, but simply to slake our curiosity about the patterns woven into the fabric of number. "The gods are there," said the modernist architect Le Corbusier, "behind the wall, at play with numbers."

To join in their play, it just takes a tiny leap of imagination.

Case in point: Here's a sort of game that's been tickling my brain recently. It begins with the so-called perfect numbers, each of which possesses a funny property: If you take the smaller numbers that divide it evenly and add them up, then you arrive back at the original number.

What's the point of this? Oh, I assure you: There is none. Despite the flattering name, perfect numbers are useless in theory, and even more useless in practice. They're a cute definition with good branding. "Perfect numbers certainly never did any good," said the mathematician John Littlewood, "but then they never did any particular harm." As my high school friend Julian used to say of pure mathematics: "Hey, at least it keeps them off the streets."

To be clear, your typical number is not perfect. Its divisors add up to less than itself (making it a "deficient" number) or more than itself (an "abundant" number).

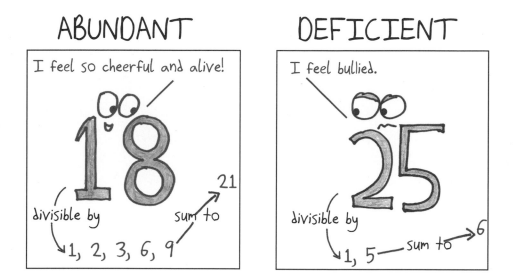

Perfect numbers are like everything perfect: elusive. The ancient Greeks knew only four examples. Ismail ibn Fallūs, a 12th-century Egyptian, found three more. By 1910, the total stood at nine. Even today, with computers so powerful we can fabricate audio of ex-presidents rapping, we have found a grand total of just 51 perfect numbers, the meager winnings of a 2,500-year Easter egg hunt.

It's hard to imagine soccer retaining its fans if only 51 goals had ever been scored.[2] So where's the fun in perfect numbers? Why play a game you almost never win?

Ah, ye of little faith. Have you already forgotten that every number is interesting?

We needn't fixate on perfection. Instead, take any old number, find its divisors, compute their sum . . . and then, starting with this new number, do it all again. If you keep going, the result is called an *aliquot* sequence.

---

2 The actual total is more than double that.

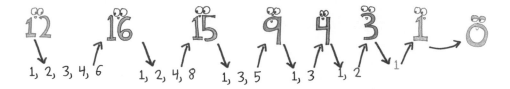

This game reveals a secret system of connections. Every number points to a new number, like an agent in a network of spies. We can pace the number line as if we were film noir detectives, sniffing out intelligence from a string of informants: 20 sends us to 22, who tips us off to 14, who brings us to 10, who suggests we try 8 . . .

SHERLOCK HOLMES AND THE GREAT ALIQUOT ADVENTURE

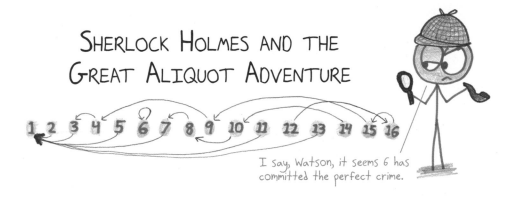

I say, Watson, it seems 6 has committed the perfect crime.

The game raises a natural question. Once an aliquot sequence is begun, where will it end?

Prime numbers, for example, all send you straight to 1 (because they have no other divisors). Perfect numbers lead to self-recommending loops: 28 sends you to 28 who sends you to 28. And some numbers form mutually recommending pairs: 220 sends you to 284, who sends you right back to 220. Such duos have an adorable name: "amicable numbers."

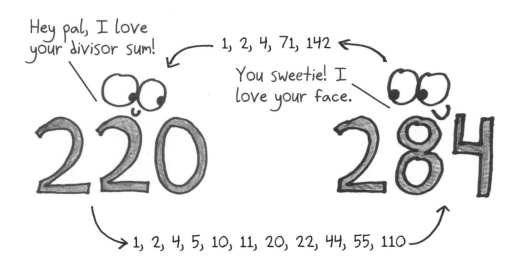

Mathematicians have spent centuries searching for these happy couples. Interestingly, the second pair (1,184 and 1,210) proved hard to spot. Big names such as Descartes, Fermat, and Euler all overlooked it, leaving it to be discovered by a 16-year-old student. Today, over a billion amicable pairs are known.

So does that cover all the ways a sequence can end? Hardly. Just as some cycles repeat after every number (e.g., $6 \rightarrow 6 \rightarrow 6 \rightarrow 6 \ldots$), and others repeat every two numbers ($1{,}184 \rightarrow 1{,}210 \rightarrow 1{,}184 \rightarrow 1{,}210 \ldots$), you can have longer cycles, which hard-core gamers call "sociable numbers." Here are two examples:

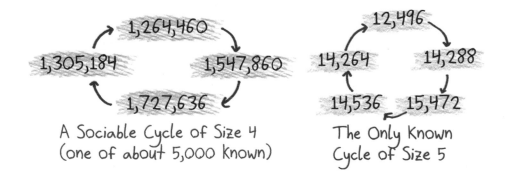

A Sociable Cycle of Size 4
(one of about 5,000 known)

The Only Known
Cycle of Size 5

When I cooked up a computer program to seek out these cycles, it uncovered a staggering conspiracy: 28 sociable numbers in a single cycle. I had stumbled upon the largest known cycle, and like countless mathematicians before me, I could scarcely believe it. What better proof that every number is interesting than to discover more than two dozen banal, unrelated civilians forming a secret cabal, the Freemasons of the integer world?

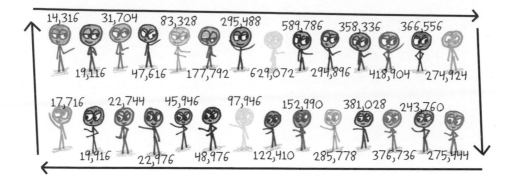

I could write a whole blathering book about this game. Did you know that some numbers (like 2 and 5) are never recommended by any others, a tragic condition known as *untouchability*? Did you catch that some numbers climb sky-high before descending, such as 138 leading to a summit of 179,931,895,322 before plummeting like Icarus back to earth? And have you wondered whether any numbers escape gravity altogether, climbing forever into the heavens? They might; we haven't figured it out. Several smallish numbers, such as 276, have unknown fates: Their aliquot sequences ascend beyond our ability to factor, like airplanes vanishing into the distance, leaving no sign of when or whether they intend to return.

Could this game one day yield practical applications? It seems unlikely. Then again, the mathematician G. H. Hardy said the same of prime numbers, which now form the basis of internet security. Though number theory begins in play, it often ends in profundity.

Perhaps, one day, our aimless game of aliquot sequences will grow into a powerful, profitable science, just as chemistry grew from alchemy, in a transformation more magical than alchemy itself. In the meantime, here are five games that invite you to romp across the landscape of number, a playground of cycles and untouchables and staggering conspiracies awaiting the next 16-year-old to detect them.

CHOPSTICKS

A FINGER GAME OF CYCLIC NUMBERS

I stumbled across this game in early 2020, became enamored, and tried to teach it to my middle school students. They responded as if I'd tried to explain high-fiving. The game wasn't just old news; it was news so ancient, I sounded unhinged calling it "news" at all. I found this to be a generational pattern. The game is old hat to folks born after 1995, and an obscurity to folks born before 1990. Like landlines, but in reverse.

How did Chopsticks sweep the schoolyards so swiftly, so completely? Well, if you need to ask, my gray-bearded friend, then you're in for a real treat.

HOW TO PLAY

What do you need? Two or more players, each with two hands. Begin like this:

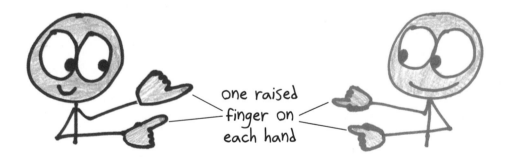

one raised finger on each hand

What's the goal? Eliminate all of your opponent's fingers.

What are the rules?

1. Take turns **tapping one of your opponent's hands** with one of yours. This leaves your hand unchanged, but **adds that many fingers to your opponent's**.

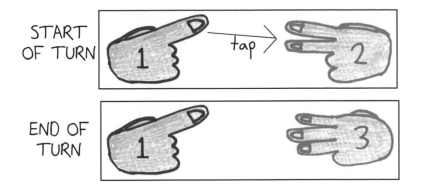

2. If you reach **five fingers, then your hand is "out,"** and resets to zero (a closed fist). This hand cannot tap or be tapped.

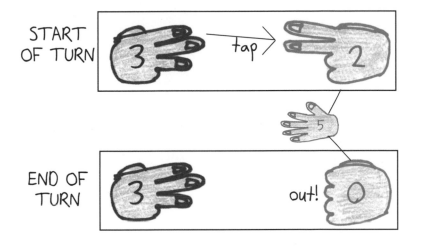

3. If a hand reaches *more* than five, then it is still in; **subtract five and continue playing**.

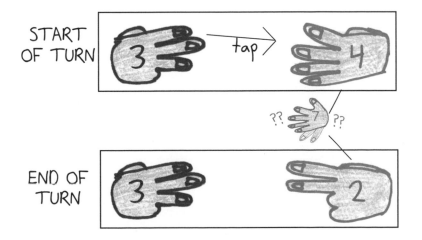

4. On your turn, instead of tapping an opponent's hand, you can **transfer fingers between your hands**. This is called a split, and may result in **reviving an eliminated hand**, or eliminating a live one.

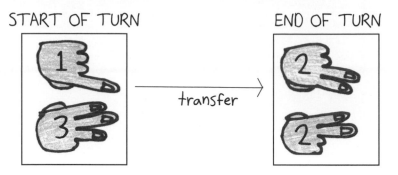

5. If **both of your hands go "out,"** then you are eliminated from the game. Last person remaining is the winner.

Chopsticks Champion

Chopsticks Chumpion

TASTING NOTES

Throughout the game, your hands are always in one of 15 states. The same is true of your opponent, which means the two-player game has 15 × 15 = 225 states in all.[3]

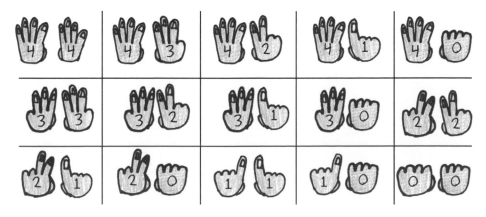

---

3   In practice, a few states turn out to be unreachable, such as both players having four fingers per hand.

With a game like this, mathematical analysis will yield one of two outcomes: (1) a surefire strategy by which one player can guarantee victory, or (2) no such strategy, meaning that sufficiently expert players will always draw.

Which kind of game is Chopsticks? As it happens, the latter. But unlike tic-tac-toe, where the board fills up after nine moves, perfectly played Chopsticks can last forever, a perpetual loop of hands tapping hands, until someone makes a mistake, or the sun engulfs the Earth, or—what amounts to the same thing—the bell rings and recess comes to an end.

WHERE IT COMES FROM

Japan, several decades ago. Beyond that, it's hard to say.

In an online survey of several generations of players (mostly from the US), only one reported learning the game before the year 2000. Oriol Ripoll, author of the delightful *Play with Us: 100 Games from Around the World*, told me the game became popular in his home of Catalonia in the early 2000s, consistent with a worldwide spread around that time.

Chopsticks goes by many names, including Finger Chess, Swords, Magic Fingers, and Split. (My students in Saint Paul, Minnesota, called it Sticks.) Some suggest that the game is about holding an actual chopstick, with the idea that raising all five fingers is a losing position because it causes you to drop the chopstick.[4]

WHY IT MATTERS

Because the anonymous children that created Chopsticks managed to reinvent one of the fundamental tools of number theory.

In the classroom, you learn that numbers never end. No matter how large your number—a billion, a trillion, a jackandjillian—you can always create something larger. But out in the rough-and-tumble world of Chopsticks, one number reigns supreme. It's a gargantuan figure, the Guinness World Record holder for largest number in existence.

I'm referring to four.

What's four plus four? Don't say eight. Chopsticks has no such concept. You might as well say four plus four equals "a handful of yesterdays" or "the ashy particles of tomorrow." Rather, as any kid from Barcelona to Kyoto can tell you, the true sum is $4 + 4 = 3$.

Confused? Don't be. Here's a handy reference table of addition facts:

---

4   I tested this theory with actual chopsticks. It checks out.

But... what happened to 8?
I swear, it was just here!

Why, that old number
burned down years ago...

Addition Table

	1	2	3	4
1	2	3	4	0
2	3	4	0	1
3	4	0	1	2
4	0	1	2	3

This isn't just an addition table. It's *the* addition table, covering every possible sum that Chopsticks allows.[5] There is nothing left to calculate. Y'all aspiring scholars should seek research problems elsewhere.

What about multiplication? Well, treating each product as repeated addition (e.g., 4 × 3 becomes 4 + 4 + 4) gives us the Utterly Comprehensive Chopsticks Multiplication Table:[6]

Multiplication Table

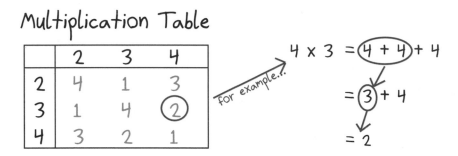

	2	3	4
2	4	1	3
3	1	4	②
4	3	2	1

for example...

$$4 \times 3 = (4 + 4) + 4$$
$$= (3) + 4$$
$$= 2$$

Mathematicians call Chopsticks math by another name: *modular arithmetic*. Or, more specifically, arithmetic mod five.

The idea is simple: Replace every number with its distance from the last multiple of 5.

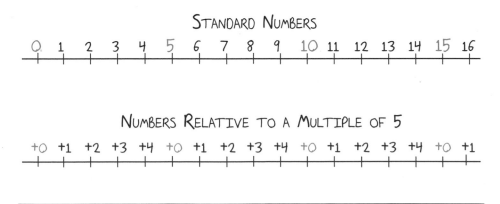

Standard Numbers

0 1 2 3 4 5 6 7 8 9 10 11 12 13 14 15 16

Numbers Relative to a Multiple of 5

+0 +1 +2 +3 +4 +0 +1 +2 +3 +4 +0 +1 +2 +3 +4 +0 +1

---

5   I've omitted those that involve adding 0, because . . . well, they're kind of boring.

6   Again, I've left out multiplication by 0 (which, as you'd guess, always gives zero) and 1 (which, as you'd expect, leaves the other number unchanged).

Chopsticks is a looping game because modular arithmetic is a looping world, a universe of endless cycles, where only five options exist. "Five more than a multiple of five"? That's just zero more than the *next* multiple. "One *less* than a multiple of five"? That's just four more than the *previous* multiple.

0, 1, 2, 3, 4: That's all you need.

Modular arithmetic pops up all over the place. For example, when asking for an International Bank Account Number (IBAN), how can I tell whether you've given me a valid one? Maybe you swapped two digits, or committed a typo, or just bashed nonsense on your keyboard in the hopes of scoring free money. It's too much trouble to keep a comprehensive list of every IBAN ever. So how does the computer know yours is real?

Easy: Any true IBAN, divided by 97, will yield a remainder of 1. Typos (or gibberish) will result in incorrect remainders. This nifty trick isn't just for IBANs: A similar process protects credit cards, national ID numbers, and even the survey codes on fast-food receipts.

Still, modular arithmetic's most audacious application is also its most familiar: tracking time.

Our clocks function on arithmetic mod 12. That is, 9 o'clock plus 7 hours equals not 16 o'clock, as a 20-fingered alien might guess, but 4 o'clock. It works on calendars, too. Have you ever seen the party trick where someone mentally calculates the day of the week for some random date that's decades in the past or the future? That stunt relies on arithmetic mod seven (because the week has seven days).

Time is an infinite-fingered creature. But for mortals like us, it's easier to imagine a time of loops, a time of cycles, a time of finite patterns (albeit in infinite repetition). We have remade time's endless game to fit our child-sized hands.

William Carlos Williams said that time is a storm in which we are all lost.

William Carlos Williams needed a better watch.

Born in a Japanese schoolyard, Chopsticks spread from continent to continent, embraced at every step by children less interested in quantifying time than in joyfully passing it. Only after the whole world had tasted the game's pleasures did grown-ups catch wise and recognize that the kids, in their playful ingenuity, had struck upon an old and fundamental truth about the cyclicity of number.

Also, not for nothing, it makes the times table way easier to memorize.

VARIATIONS AND RELATED GAMES

CHOPSTICKS MOD N: Play as if you have a different number of fingers, such as 6, 7, or 99. You'll want to use pencil and paper!

CUTOFF: Instead of using modular arithmetic, any hand with more than five fingers is immediately declared "out."

MISÈRE: The game in reverse; you win by having both of your hands go out.

ONE-FINGERED DEFEAT: If one of your hands is "out," and the other has only one finger remaining, then you lose. Having both hands go out still counts as a loss, too.

SUNS: Begin with four fingers on each hand, instead of just one. Interestingly, your game will never return to this starting position.

ZOMBIES: If you are eliminated in a multiplayer game, then you continue to play with a single one-fingered hand. On your turn, you may tap an opponent, but no one may ever tap you.

SEQUENCIUM

A GAME OF RIVAL VINES

Of all the games I play-tested for this book, Sequencium received some of the most rapturous reviews. Perhaps that's because its numbers act like none you've ever seen. Rather than stand in tidy formations, they snake and slither across the board, like the thirsty tendrils of a sentient plant. I'm not surprised that Walter Joris, who has designed hundreds of games, considers Sequencium his masterpiece.

HOW TO PLAY

What do you need? Two players, each with a different color pen, and a 6-by-6 grid. For a longer game, try an 8-by-8 grid (or 7-by-7 grid with the middle square blacked out).[7] Each player begins with the numbers 1, 2, and 3 on the grid as shown.

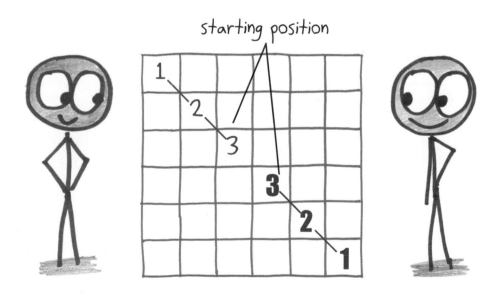

What's the goal? Reach a higher maximum number than your opponent does.

What are the rules?

---

7 On odd-sized boards, taking the middle square would give Player 1 a big advantage. We black it out to keep the game fair.

1. On each turn, (1) **pick one of your existing numbers**, (2) **add one** to it, and (3) write the **new number in an adjacent cell**. Diagonals count as adjacent.

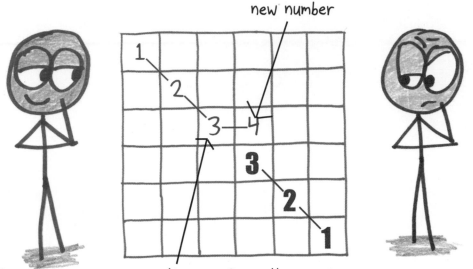

2. You may **build off of any of your existing numbers**, as long as there is space to do so. Also, it's **okay to cross an existing path** along a diagonal.

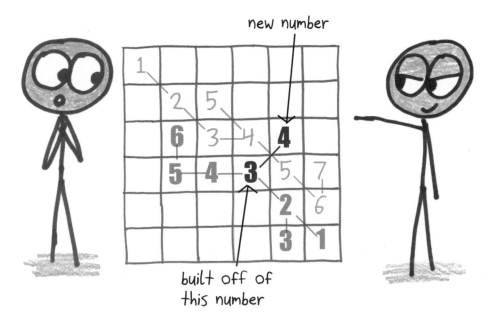

3. **Play until the board is filled**, even if one player becomes unable to move.

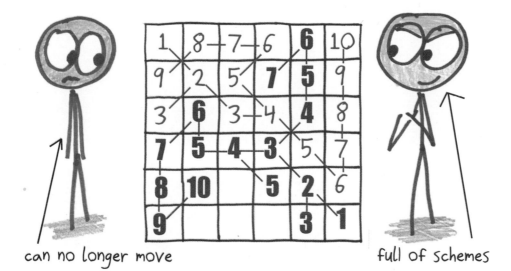

can no longer move

full of schemes

4. The winner is whoever, in the end, has written the **highest number**.

winner!

Experts may wish to adopt the rule change described under "Tasting Notes."

TASTING NOTES

I love this game. So let me go ahead and spoil it for you.

Like many games of pure strategy, Sequencium favors the first player. That doesn't necessarily spell doom; after all, people seem to enjoy chess, even though the first mover wins about 55% of the time. However, unlike chess, Sequencium affords the second player a game-killing response: Just copy the first player. Rotating the board 180° and playing symmetrically will guarantee a draw.

To nullify that stratagem, I advise a simple tweak. The first player's opening move unfolds as usual. Then, starting with the second player's first move, **each player moves twice per turn**.

"Break the symmetry," notes game designer and play tester Joe Kisenwether, "and you have a gem of a game. Do you block off territory of your own, or concentrate on cutting off your opponent? Do you 'stick with them,' or try to push them out of your space?

"It feels like a classic," says Joe. "How is it that this game isn't ancient?"

WHERE IT COMES FROM

Despite its timeless feel, Sequencium was born in the 21st century, from the strange and fertile mind of Walter Joris. In the years since publishing his book *100 Strategic Games*, he has continued to churn out puzzles, origami designs, bizarre art projects, unsettling cartoons, and new games (Sequencium among them). His stuff is so weird and compelling that I consider him a kind of human pulsar, emitting what I can only call Joris radiation.

Anyway, when I asked him his favorite game of all those he has designed, Walter didn't hesitate. Sequencium, though previously unpublished, is his crown jewel.

WHY IT MATTERS

Because in designing a fair system, there's no greater challenge than how to take turns.

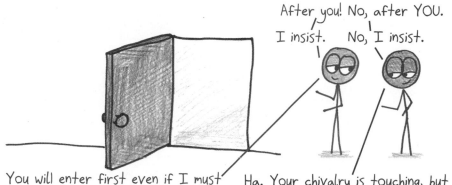

After you! No, after YOU.
I insist.   No, I insist.

You will enter first even if I must   Ha. Your chivalry is touching, but
fling your unconscious body through.   I'd sooner murder you than relent.

To witness the standard approach to turn-taking, just stop by a schoolyard at recess and watch two captains picking teams.[8] First you pick, then I pick, then you pick, then I pick, then you pick . . . and so it goes, until even the slimmest of pickings has been duly picked.

This process is simple, easy, and grossly unfair.

You begin with a distinct advantage: the #1 pick vs. my inferior #2. Then, on the very next turn, you gain another advantage: the #3 pick vs. my measly #4. Before I can lodge a formal complaint, you surge further ahead, claiming the #5 pick while I'm left with paltry #6.

These tiny advantages accumulate into a large one, known as *first-player advantage*. It looms like a storm cloud over the world of games, a perpetual threat to fairness.

---

8   Or don't. Schools haven't embraced the idea of live audiences for recess sports.

Take chess. For clumsy novices like me, it's balanced enough. But for top players, second-mover Black and first-mover White are as distinct as . . . well, as black and white. "The tasks . . . are different," wrote grandmaster Evgeny Sveshnikov. "White has to strive for a win, Black—for a draw!" The first mover can attack freely, while the second mover begins on the defensive. "When I am White I win because I am White," Efim Bogoljubov once said. "When I am Black I win because I am Bogoljubov." Gotta love his energy.

I could go on. First-player advantage nibbles like a termite at the foundations of Connect Four, Monopoly, Risk, Hex, checkers, go (where experts quantify first-mover advantage as worth roughly 6 to 7 points), and Sequencium, among countless others. But why ruminate on injustice, as if justice lay beyond our grasp?

Why not bust out some righteous mathematics instead?

Questions of resource allocation—even of an intangible resource, like turns in gameplay—are inherently numerical. It's no surprise that our search for fairness drives us into math's cold and impartial arms.

In his book *New Rules for Classic Games*, R. Wayne Schmittberger collates several clever systems for neutralizing first-player advantage. First, a free-market solution: **Let players bid for the right to go first**. In Sequencium, for example, I might say, "Let me go first, and I'll add 1 to your final score." Then, you can either raise ("Let me go first, and I'll add 2 to your final score") or accept my bid.

System #1: Bidding

Second, a meta solution: **Play two games, one in each role, and add together the scores**. Sounds fair enough, but in an ironic twist, this method may confer an advantage on the *second* player. (That is, the player whose chance to go first comes in the second game.) Entering the latter game with a clear target to aim for, they can adjust strategy accordingly.[9]

---

9   Schmittberger suggests a wise (if dizzying) fix: play both games simultaneously.

System #2: Trade Roles

Third, a classic mathematical solution, known as the pie rule or **"I cut, you choose."** The idea comes from the dessert table. One person cuts the treat in two, and the other person chooses their preferred half. The cutter, knowing they'll get the smaller piece, will strive for perfect equality. To apply the procedure to Sequencium, I make the first moves for both sides; then, you decide which side you want to play.

System #3: "I Cut, You Choose"

All of these are clever ideas. But to balance Sequencium, I favor a fourth method: Change the very meaning of "taking turns."

It sounds radical. But is it? No stone tablet or burning bush has mandated that players always alternate back and forth. Fantasy sports leagues, for example, often run a "snake" draft. A picks, then B, then C, followed by another turn for C, then B, then A, then another turn for A, then B, then C, and so on. As the turns elapse, the role of "first player" alternates. The first shall be last, and the last shall be first.

In Sequencium, this method works like a charm. With each double move, you can respond to your opponent's last sally, and then launch an attack of your own.

To be fair, even this scheme suffers a flaw. We're essentially taking turns at having first-player advantage. First you experience it; then I do; then you; then me. But that gives you a first-player advantage *in experiencing first-player advantage*. Whether it's our first, seventh, or ninety-third chance to taste that advantage, your taste always comes just before mine.

New level of abstraction, same old problem.

Will this upset the balance of a real-world game? Probably not. But it may upset your friendly neighborhood mathematician. Luckily, there's a more robust solution, a system of turn-taking that ensures perfect balance at *every* level of abstraction.

To begin, you take a turn (marked with "0"), then I take a turn (marked with "1").

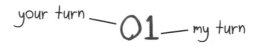

After that, we consider the sequence of turns thus far, copy it, swap the roles of "you" and "me," and carry out those turns next.

first 2 turns—01
10—next 2 turns

Then we repeat the process: copy the sequence thus far, swap the roles of "you" and "me," and carry out those turns.

first 4 turns—0110
1001—next 4 turns

Then we do it again.

first 8 turns—01101001
10010110—next 8 turns

And again.

first 16 turns—0110100110010110   next 16
1001011001101001—turns

And again, until the game is over.

0110100110010110100101100110100 1
1001011001101001011010011001 0110

This decadent slice of mathematics—developed by a number theorist, rediscovered by a chess master, and known today as the Thue-Morse sequence—offers the fairest alternation of turns imaginable. It ensures balance not only in first-player advantage (because 0 precedes 1 as often as 1 precedes 0), but in the advantage of wielding first-player advantage (because 01 precedes 10 as often as 10 precedes 01), and in the advantage of wielding *that* advantage (because 0110 precedes 1001 as often as 1001 precedes 0110), and so on, ad infinitum. A book on the mathematics of fair division dubbed it "taking turns taking turns taking turns . . ."

Today, wherever turns are to be taken—penalty shoot-outs, tennis tiebreakers, the sharing of bites at an Ethiopian restaurant—you'll find mathematicians smiling a little too intensely as they force the Thue-Morse sequence into the hands of unsuspecting civilians.

Case in point: Ever noticed that the liquid at the bottom of the coffeepot is more potent than the stuff at the top? Well, to create two equally strong cups of joe, pour out tiny amounts in the style of Thue-Morse. A splash for the left mug, a splash for the right, a splash for the right, a splash for the left, and so on, until your coffee is so cold you need to brew a new pot anyway.

Though it has some genuine applications, I offer up the Thue-Morse sequence mostly as a playful abstraction, an illustrative ideal, showing that even a silly game like Sequencium can help reveal the theoretical structure of perfect fairness. The road to higher understanding is paved by play.

By the way, if you want to play Thue-Morse Sequencium, I recommend charting the upcoming turns with pencil and paper.

I also recommend not despairing when you inevitably lose track.

VARIATIONS AND RELATED GAMES

THREE PLAYERS: Use a triangular board, and a "snake draft" order of turns. Any triangles sharing a corner count as adjacent.[10] The board depicted below is good for a warm-up, but for a richer game, I recommend adding another few rows of triangles.

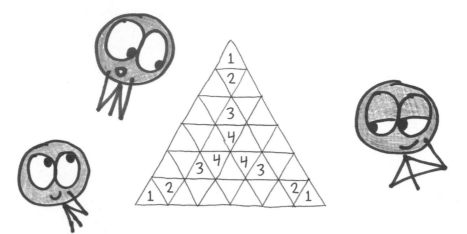

FOUR PLAYERS: Play on larger grid (8 by 8 or 10 by 10), with each player's chain beginning in a different corner, and a "snake draft" order of turns.

FREE START: Begin with a blank board, and allow players to place their connected trio of initial numbers anywhere that they like. (Thanks to Mihai Maruseac for the suggestion.)

FRESH SEEDS: At any time, you may play a 1 in any open square, even if it is not touching any of your own numbers. This counts as *both* of your moves for that turn. (Thanks to Andy Juell for the suggestion.)

STATIC DIAGONALS: This is an interesting rule change proposed by Katie McDermott, to decrease the power of diagonal moves. When playing horizontally or vertically, your numbers increase as usual. But when playing diagonally, the number remains the same. If using this rule, you may want to begin the game with just a 1 in opposite corners, rather than the usual 1-2-3 formation.

---

10 A hexagonal board doesn't work as well, because most of the game's surprises come from "diagonal" moves between squares connected only by a corner, and the hexagonal board has no such connections.

33 TO 99

ADD, SUBTRACT, MULTIPLY, DIVIDE . . . AND OCCASIONALLY RAGE-QUIT

A father and son are lounging on the living room floor, bonding as only kinsfolk can: by watching YouTube videos of trains. Then, noticing the clock, the son leaps up and retrieves his math homework. It consists of a single, seemingly straightforward question.

Father scratches his head. Father frowns. Father sends Son away on a pointless errand. Then, with the coast clear, Father whips out his Nexus 7 tablet and googles the answer.

Okay, yeah, it's a tablet commercial. But it's a tablet commercial with 3 million views on YouTube, a tablet commercial that launched a viral sensation in Japan, because the problem is trickier than its grade-school surface. Answers like $8 + 5 - 1 - 1$ and $\frac{8}{1+1} + 5$ don't work, because you've got to hit 10 exactly. Meanwhile, you can't do $1 + 1 + 8$, or $5 \times (1 + 1)$, because you've got to use all four numbers.

Give it a shot. For those curious, I'll bury the answer in a footnote.[11]

In the meantime, this chapter extends such puzzles into a game that may challenge your conception of what "simple math" means—and of who is likely to excel at it.

---

11 No, not *this* footnote.

HOW TO PLAY

What do you need? Two to five players (though more is fine), each with pencil and paper. Also, five standard dice (easily simulated; search the internet for "roll dice") and a timer. I recommend 1- or 2-minute rounds, but preferences vary, and the game can work without any time pressure at all.

What's the goal? Get as close as you can to the target number, without going over.

What are the rules?

1. One player, the leader for the round, **calls out a target number**, anywhere from 33 to 99. Then the leader **rolls the five dice** and starts the timer. (Again, feel free to skip the time pressure if that's not your jam.)

I choose a target number of... fifty-five!

2.   Every player tries to **achieve the target number by combining the five dice via the four operations** of addition, subtraction, multiplication, and division. You must use each of the dice precisely once, but you can pick and choose (and repeat) operations. Parentheses are allowed. Your final answer must be equal to or less than the target, and it must be a whole number, though fractions are permissible as middle steps.

3.   When the timer goes off, compare your results. **Your score is your distance from the target.** (Thus, lower scores are better.) To keep things civil, the **maximum score for a round is 10**.

4.   Play until everyone has an equal number of turns calling the target. In the end, the person with the **fewest points wins.**

TASTING NOTES

You know how movies show ephemeral, glowing numbers swirling around the head of anyone who's thinking hard about math?

Ever craved that experience in real life?

I'm not saying 33 to 99 will turn you into a cinematic math genius. But it may be the next best thing. Grab a pencil and paper, and watch as the numbers dance and spiral, forming unsuccessful combinations and then separating to try again. Even when I'm struggling, I feel like I'm flying.

WHERE IT COMES FROM

The core idea goes back centuries. In the 1700s, textbooks featured puzzles such as "I can place four 1s so that, when added, they shall make precisely 12. Can you do so too?"[12] In 1881, the famed Four Fours puzzle was first published, challenging readers to hit every target number from 1 to 100 by using four 4s. (It requires some creative uses, such as 44, $\sqrt{4}$, 4!, and .4). In the 1960s, the 24 Game (in which the target is always 24) spread through Shanghai and other cities in China. Finally, a version of the game began airing on French TV in 1972, and British TV (under the name *Countdown*) a few years later. I found this particular ruleset in the book *Dice Games Properly Explained*, by game designer Reiner Knizia. He calls it Ninety-Nine.

WHY IT MATTERS

Because when math opens the door a little wider, you never know who will enter and thrive.

You probably know what it feels like in a typical math classroom. Day after day, the same students finish first, the same students lag behind, and the same students doodle crocodiles eating their wrong answers. Watching a teacher return a graded test, you'll taste the atmosphere of a tournament: winners vs. losers, A's vs. F's, "math people" vs. "not math people."

I'm here to tell you that it doesn't have to be this way.

Minnesota teacher Jane Kostik once introduced the 24 game (a variant of 33 to 99) to a remedial high school class. Her goals were modest: just to support their shaky arithmetic skills. But the game hooked them. Racing through combinations got their blood pumping in a way that closed-form questions, with a single path to the answer, never did. Their whooping and applause got so loud that the calculus students across the hall would stand in the doorway to watch. "Eventually," Jane told me, "the calculus class challenged my class to a competition."

---

12 The given answer was 11 + ¹⁄₁, but there are others.

Schools sort kids by mathematical success. A class of the most successful had just walked into a class of the least successful and declared war. It's as if the varsity hockey team decided to take on a ragtag bunch of kids who had tried out and missed the cut.

Yet, in the underdog tradition of the finest sports movies, Jane's class won.

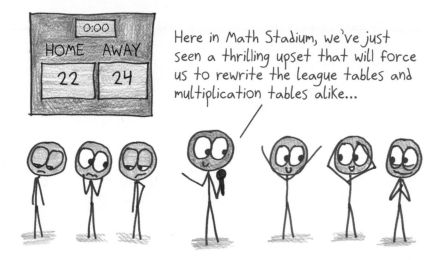

Most math problems follow a familiar format: "Here's the computation; what's the result?" 33 to 99 flips the script: "Here's the result; what's the computation?" Because five numbers can be combined via the four operations in thousands of ways, you can never check all the possibilities. That leaves the door open for intuition, creativity, and the occasional flash of genius.

Not to mention bandwagons full of people you'd never have pegged as math lovers.

Take the British TV show *Countdown*, half of which is spent playing a variant of 33 to 99. This seemingly niche pastime has aired over 7,000 episodes. The editor-in-chief of Guinness World Records called it a "cornerstone of British popular culture," putting it on par with witty rejoinders, goopy desserts, and not pronouncing your *r*'s.

Here's a typical puzzle, this one from 2010. Two contestants strive to combine six numbers to hit a lofty target:[13]

---

13 Two key rule changes here: (1) You don't need to use all the numbers, and (2) going over is fine; whether you wind up above or below the target, all that matters is how close you get.

After 30 seconds, one contestant shrugs in defeat, while the other announces that he came close:

"Excellent," says the host. "Can it be bettered, Rachel?"

At this point, Rachel Riley—the telegenic presenter of the game, who has been transcribing the contestant's operations—casually remarks, "Yeah, this one is possible." Then she tosses off this tour de force:

Like her *Countdown* predecessor Carol Vorderman, Rachel Riley is a peculiar figure: a tabloid-hounded celebrity who built her career on quick mental math. Her professional duties are (1) be beautiful, (2) be charming, and (3) solve intricate arithmetic problems at blinding speed on national television. I myself mastered #1 and #2 long ago, but I'd need years to approach Rachel's virtuosity on #3.

Now, if you're a stodgy and overeducated sort, you might argue that none of this is "real" math. It's a running joke among mathematicians that they're bad with numbers. Though this confession confuses outsiders—it's like hearing surgeons plead clumsiness, or poets boast of illiteracy, or Rick Astley declare that he *is* going to give you up and let you down—the point is that mathematicians aren't professional arithmetic doers, any more than musicians are professional instrument tuners. Mathematics is a craft of abstraction and problem-solving, not calculating tips.

In that sense, though, games like 33 to 99 actually strike closer to the true nature of math than a standard worksheet. Any fool with a calculator can compute $(10 + 6 + 1) \times 37 + (5 \times 4)$. But to explore different combinations of 1, 4, 5, 6, and 10 in search of a total that when multiplied by 37 yields a result whose difference from the target of 649 is precisely equal (deep breath) to a combination of the remaining numbers?

That takes strategy. It takes skill. It takes no small measure of inspiration. Or at least, it takes a Nexus 7 tablet.[14]

VARIATIONS AND RELATED GAMES

THE 24 GAME: A high-pressure, race-the-clock version of 33 to 99.

1. Set the timer to 24 seconds, but don't start it yet.
2. Roll four dice (ideally 10-sided, but 6-sided is fine). Everyone tries to combine the dice to achieve the target number of 24, using all four dice, and any operations you wish.
3. Whoever manages it first shouts "Twenty-four!" and then starts the timer. Everyone else has 24 seconds to find a solution.
4. Anyone who finds the solution before time expires (including the person who originally found it) scores 1 point. Mistakenly calling "Twenty-four!" when you don't have a solution costs you 1 point. First to 5 points wins the game.

BANKER: A version of 33 to 99, also from Knizia's *Dice Games Properly Explained*. This one requires only a single die, yet has a bigger role for chance. It also adds the constraint that you must use each operation precisely once.

Here's how each turn goes:

1. Roll the die and mark down its value.
2. Roll again. Take your previous value, and add, subtract, multiply, or divide by the new number to arrive at a new value. (When dividing, ignore remainders.)
3. Continue doing this until you have used all four operations once each. You cannot repeat any operations. Your final value is your score for the round.

Play for an agreed-upon number of turns. The highest total score wins.

---

14 I almost forgot! The solution to the puzzle from the Japanese commercial is a doozy: $\frac{8}{(1 - \frac{1}{5})}$. You can check for yourself: the denominator gives ⅘, or 0.8, which goes into 8 precisely 10 times.

Roll #1 Score: 4.

Roll #2 I'll divide. Score: 2.

Roll #3 I'll add. Score: 8.

Roll #4 I'll multiply. Score: 24.

Roll #5 I must subtract. Final Score: 23.

NUMBER BOXES: This game, popularized by educator Marilyn Burns, turns 33 to 99 inside out. Here, you have no control over the operations; instead, you get to control the order in which the numbers are combined.

To begin, each player draws a blank copy of the same calculation, including extra "throwaway" boxes off to the side. Perhaps something like this:

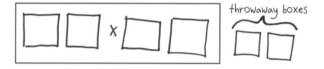

Then, someone rolls a die (ideally 10-sided, but 6-sided is fine), and every player writes the resulting number in a blank spot. You're allowed to skip two numbers by placing them in your throwaway boxes.

After every box is filled, carry out the calculation. The winner is whoever comes closest to a predetermined target (say, 2,500).

Math educator Jenna Laib calls this game "the ultimate chameleon" because with the right choice of calculation it can suit any level of mathematics.

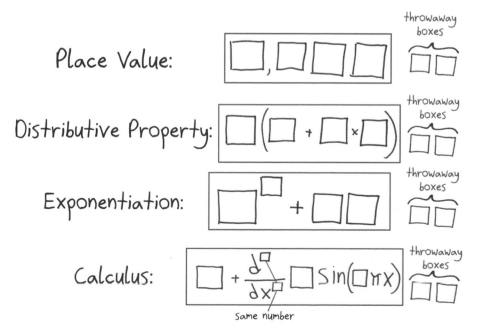

Place Value:

Distributive Property:

Exponentiation:

Calculus:

same number

PENNYWISE

A GAME OF MAKING CHANGE

Don't take this the wrong way, US penny, but you're worthless. Actually, it's worse than that: since you cost the US Treasury more than your own value to mint, you are *less* than worthless. You are a negative number made of zinc. We'd be better off forcing you into retirement and rounding prices to the nearest 5¢.

Except in Pennywise. Here, your dull copper coat gets a chance to shine. You supply not only the game's title, but also its core strategy. "Save your pennies," advises the game's creator, James Ernest. And so, penny, we shall save you—and when the time comes, we ask that you in turn save us.

HOW TO PLAY

What do you need? Two to six players and a jar of coins.

What's the goal? To be the last one with coins remaining.

What are the rules?

1. Each player begins with **four pennies, three nickels, two dimes,** and **one quarter**.

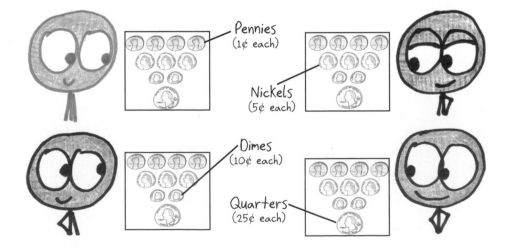

Pennies
(1¢ each)

Nickels
(5¢ each)

Dimes
(10¢ each)

Quarters
(25¢ each)

2. On each turn, **place one of your coins in the center of the table**. Then, you may **take back any combination of coins whose total value is strictly less** than the coin you put in. For example, if you put in a dime (10¢), you can take back at most 9¢ of change. On some moves (including, notably, the first move of the game), you **will not be able to take back any change at all**.

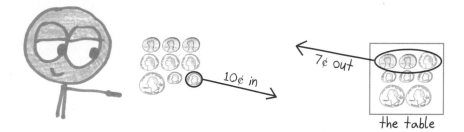

3. **Last player with any coins remaining is the winner**.

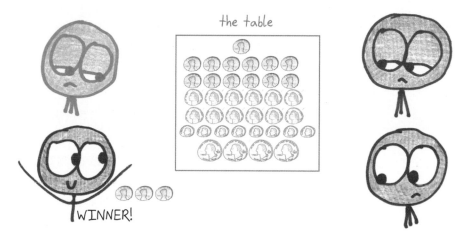

TASTING NOTES

The first imperative is clear: Take as much change as you can. Since you start with 64¢, and lose at least 1¢ each step, you can in theory remain solvent for 64 turns. But every time you take back less than the maximum—say, 6¢ for a dime, or 15¢ for a quarter—your turns dribble away. Worst-case scenario, taking no change at all, you'll go bankrupt in just 10 moves.

Each penny buys you precisely one turn: You put it in and take nothing out. Meanwhile, a nickel *might* buy you five turns, but only in the rare scenario where you can take four pennies as change. If no pennies are available, then the nickel buys you a single turn. The same pattern holds for dimes and quarters. Indeed, the higher a coin's value, the less likely that you can milk it for the maximum number of turns.

This points to a possible strategy: Ditch your pennies first, and save your dimes and quarters for when you can make proper change. Should the game be renamed Quarterwise?

Nope. Try that approach, and your opponents will snatch up the pennies and nickels like Wall Street vultures, leaving you to make shoddy change anyway. Instead, the game demands a careful trade-off. Minimize your losses each turn, but don't fork over the small coins too soon.

Who knew making change required such sophistication?

WHERE IT COMES FROM

The game was devised by James Ernest, founder and proprietor of Cheapass Games. "Pennywise is one of our oldest titles," he writes, "and one of the simplest." At one point, he even printed it on the back of the company's business card.

The game also calls to mind a whole genre of puzzles about making change using as few coins as possible.

For example, using common US coinage, it requires a minimum of nine coins to make 99¢: three quarters, two dimes, and four pennies. ($1 is easier; you can just use four quarters.) The puzzle: What amount *below* 99¢ requires the most coins, and how many coins does it require?[15]

Other coinage puzzles play with the system itself. In the US, making every amount from 1¢ to 99¢ requires a total of 470 coins: one for 1¢, two for 2¢, and so on, all the way up to nine for 99¢. Can we reduce this number by changing the denominations—that is, by abandoning the penny-nickel-dime-quarter system in favor of some other collection of four values?

As it turns out, we can. Just replacing the 5¢ nickel with a 3¢ coin (trickle?) shaves 50 coins off the total required. And some offbeat combinations—like coins worth 1¢, 4¢, 11¢, and 39¢—perform even better. Still, I'm grateful the US Treasury hasn't adopted them, lest every candy bar purchase bring the world economy grinding to a halt.

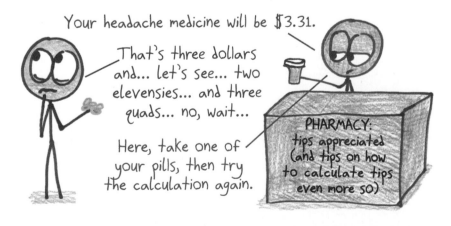

---

15 Spoilers: Making 94¢ also requires 9 coins (3 quarters, 1 dime, 1 nickel, and 4 pennies).

A good challenge for the skillful computer programmer: What four denominations allow you to make every value from 1¢ to 99¢ with the fewest total coins?

WHY IT MATTERS

Because it is the bedrock of civilization.

I don't mean to overpraise the game. You could ditch Pennywise and, with a few lucky breaks, probably retain a viable society. All I mean is that Pennywise animates the mathematical concepts that make our economic life possible.

First, *simple tokens*. Ancient societies across the globe tracked their possessions—sheep, goats, Super Bowl rings—with basic clay tokens, each standing for a single good. One token, one sheep. Two tokens, two sheep. Three tokens, three sheep. Four tokens . . . and uh-oh, now you're sleeping. Such are the dangers of sheep tabulation.

This system of one-to-one correspondence is perhaps how civilization itself first understood numbers.

Next came *complex tokens*. Whereas ancient societies in China and Mesoamerica satisfied themselves with simple tokens, the folks in Sumer hatched a new idea: one token to stand for *multiple* sheep. These were the nickels, dimes, and quarters of the livestock economy.

Third, *abstraction*. Prior to this step, there is no free-floating symbol for 6. Numbers are expressed only in connection with a particular quantity, with each quantity represented by a single symbol: "seven sheep," "seven goats," "seven Super Bowl rings for the GOAT." The number 7 cannot be separated from the good that it enumerates.

But over time, "three sheep" becomes denoted by two symbols, as does "three goats," with a shared symbol for 3. Thus arrives the decisive feature of mathematics: an abstract concept of "number."

Not three sheep. Not three goats. Not three anythings. Just "three."

The origin story of number has one final twist. The Sumerians soon began placing their tokens inside clay envelopes, and then, to indicate the envelopes' contents, they marked them by pressing symbols into the wet clay. Archaeologist Denise Schmandt-Besserat has argued that this practice gave birth not only to Sumerian mathematics, but also to something just as profound: Sumerian writing. We owe the human tradition of literacy, in no small part, to the bookkeeping of sheep.[16]

I'm simplifying, of course. This wasn't the only pathway that led to writing. Also, complex tokens weren't currency as we'd recognize them, because they weren't used for exchange. They were more like ledgers, or bank accounts: not coins, but records of ownership.

Still, every monied society today owes those tokens a tip of the hat. Or, at the very least, a penny in the jar.

---

16 You may call this "booksheeping" or "sheepkeeping." Your pick.

VARIATIONS AND RELATED GAMES

OTHER STARTING COINAGE: Rather than beginning with the coins mentioned above, try out these alternatives (each proposed by James Ernest, the game's creator).

Name	Coins at the Beginning	Coins	Total
Classic	1, 1, 1, 1, 5, 5, 5, 10, 10, 25	10	64¢
Coprimes	1, 1, 1, 1, 4, 4, 4, 7, 7, 13	10	43¢
Darlene	1, 1, 1, 3, 3, 3, 10, 10, 20	9	52¢
No Dimes	1, 1, 1, 1, 5, 5, 5, 25	8	44¢
Sugar	1, 1, 2, 2, 5, 5, 10	7	26¢
Taylor	1, 1, 1, 5, 5, 10	6	23¢

Or you can use these denominations, inspired by real national currencies:

Country	Coins	Coins	Value	Comments
Djibouti	1, 1, 1, 2, 2, 2, 5, 5, 10	9	29¢	The US is missing out on the fun here; 2-unit coins are common worldwide.
Chile	1, 1, 1, 1, 5, 5, 5, 10, 10, 50	10	89¢	Most currencies have a 20¢ or 25¢ coin. Only a handful have neither.
Bhutan	1, 1, 1, 1, 5, 5, 5, 10, 20, 25	10	74¢	Most currencies have a 20¢ or 25¢ coin. Very few have both.
Azerbaijan	1, 1, 1, 1, 3, 3, 3, 5, 5, 10	10	33¢	Along with Cuba and Kyrgyzstan, one of the few countries with a 3-unit coin.
Madagascar	1, 1, 1, 2, 2, 2, 4, 4, 4, 5, 5, 10	12	41¢	The only country I found with a 4-unit coin. Sparkle on, Madagascar!

Of course, you can also play around with your own combinations. Whatever you pick, make sure all players begin with identical stashes.

NEW CHANGE-MAKING RULES: These two interesting variations come from Joe Kisenwether.

1. *Perfect Change:* You may take back any combination of coins whose value is less than **or equal to** the coin you put in, as long as they are of **a lower denomination**. For example, if you put in a dime, you can take back two nickels, though you cannot take back a dime.

2. *More-than-Perfect Change:* You may take back **all the coins of a lower denomination than the one you put in, even if their total value exceeds your original coin**. For example, you could put in a dime, then take out three nickels and five pennies.

A good puzzle from Joe: Could either of these variations lead to an endless game? If not, what is the maximum number of turns that the game could last?

FLIP: A two-player dice game from James Ernest. To begin a round, each player **rolls five standard dice**. The player with the lowest total goes first. Then, on each turn, you may either:

1. ***Tap one of your opponent's dice.*** They must now place that die in the center of the table, and in return, they may take from the middle any combination of dice whose total is strictly less than the die they just put in. For example, if you lose a 5, you could take any dice whose sum is 4 or lower. (Sometimes no such dice are available.)

2. ***Flip over one of your own dice.*** The two numbers on opposite sides of a die always add to 7; thus, flipping turns 1s into 6s, 2s into 5s, and 3s into 4s (and vice versa).[17]

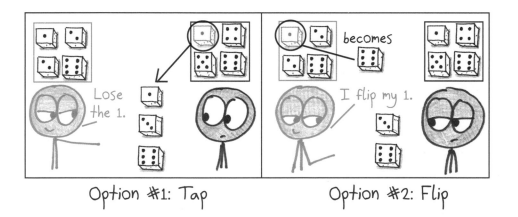

Option #1: Tap   Option #2: Flip

The **last player with dice remaining is the winner of the round** and scores the total of their remaining dice. First to reach a score of 50 wins the game.

---

17 One caveat: Once you flip a particular die, you cannot flip it again until after you've tapped one of your opponent's dice. This rule is necessary to avoid stalemates.

PROPHECIES

A GAME OF SELF-FULFILLING (AND SELF-DEFEATING) PREDICTIONS

A prophecy isn't just a prediction. It's an action. It can even, in a headache-inducing way, alter the very future that it aims to forecast.

For example, you can't say to an expecting couple "your baby is going to kill papa and marry mama" without wreaking some rather drastic changes in their parenting style. Similarly, you can't tell a credulous audience of millions "I expect this stock to go sky-high!" without causing extra purchases, thereby driving up the price. And before you inform an evil wizard Who Shall Not Be Named that a baby wizard (astrological sign: Leo) may cause his downfall, consider the causal implications of your own prophetic words.

All of this lies in the background of Prophecies, an elegant case study in self-fulfilling (and self-defeating) predictions.

HOW TO PLAY

What do you need? Two players, each using a different color pen. Also, paper with a rectangular grid (square is perhaps preferable, but not necessary) with four to eight rows, and four to eight columns.

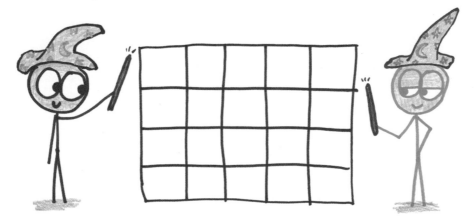

What's the goal? Accurately predict how many numbers will appear in a given row or column, by writing that number somewhere in the row or column.

What are the rules?

1. Take turns **marking empty cells with either a number or an X**.

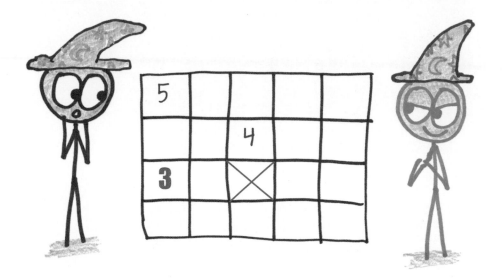

2. Each number is a kind of prophecy: a **prediction of how many numbers will eventually appear in that row or column**. Thus, the smallest usable number is 1, and the largest is the length of the row or column (whichever is larger). Meanwhile, an **X simply fills up a spot**, ensuring no number appears there.

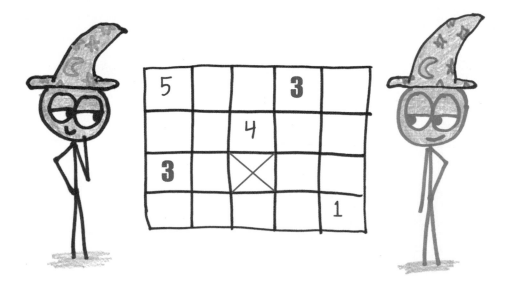

3. To avoid repeat prophecies, **no number can appear twice in a given row or column**.

4. **If a cell becomes impossible to fill**, because any number would be a repeat prophecy, **mark it with an X**. This is just a friendly act of bookkeeping; it does not count as a turn.

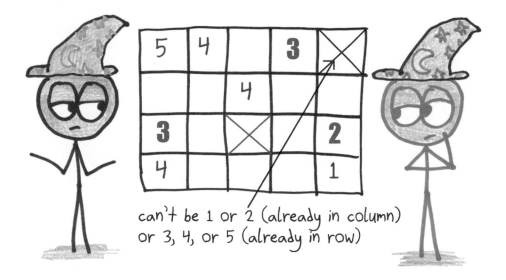

5. Play until the board is full. Then, count the numbers appearing in each row. **Whoever made the correct prophecy in that row is awarded that number of points.** Then do the same for the columns.

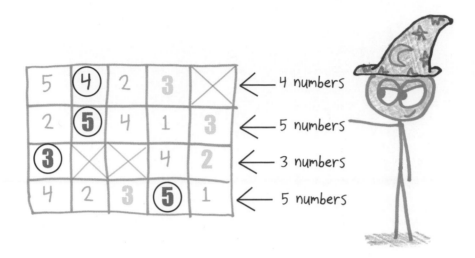

Note that **a single prophecy may score twice**: once in its row, and once in its column. Meanwhile, **some rows or columns may contain no correct prophecy.**

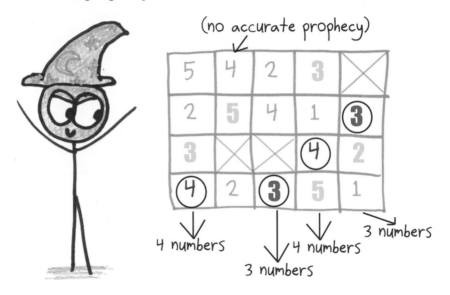

6. **Whoever scores more points is the winner.**

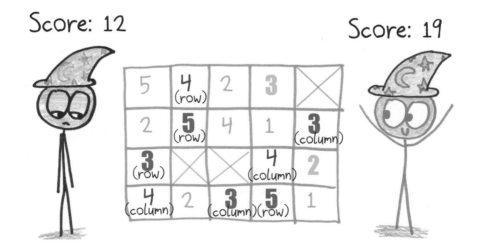

TASTING NOTES

The game's most delicious and agonizing moments occur when a prophecy threatens to undermine itself. Take the following row:

It's guaranteed to wind up with two or three numbers. Thus, with 3 already claimed, Green would love to place 2—except that introducing this number would be a pointless act, negating itself and validating the rival 3. It's a classic conundrum for prophets: To describe the world's contents, you must alter them, thereby rendering your description false.

Such moments remind me of self-enumerating sentences. These are linguistic inventories, not of any outside object, but of themselves. Consider the first published example:

"Only the fool would take trouble to verify that his sentence was composed of ten a's, three b's, four c's, four d's, forty-six e's, sixteen f's, four g's, thirteen h's, fifteen i's, two k's, nine l's, four m's, twenty-five n's, twenty-four o's, five p's, sixteen r's, forty-one s's, thirty-seven t's, ten u's, eight v's, eight w's, four x's, eleven y's, twenty-seven commas, twenty-three apostrophes, seven hyphens and, last but not least, a single !"

—Lee Sallows

The sentence accurately describes itself. Which is wild. Trying to imagine how Lee composed it twists my brain into pretzel loops, because every tweak to the sentence must anticipate its own consequences.

To see what I mean, try completing this simple self-describing chart:

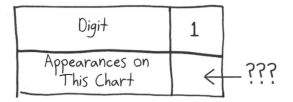

So far, there is a single 1. But when we try to fill out 1 in the second row, we get tangled in our own shoelaces, as another 1 appears. Perhaps we ought to replace it with a 2. But the moment we do so, the second 1 vanishes, and our 2 is thereby rendered false.

Any prophecy here undermines itself. The table is unfillable.

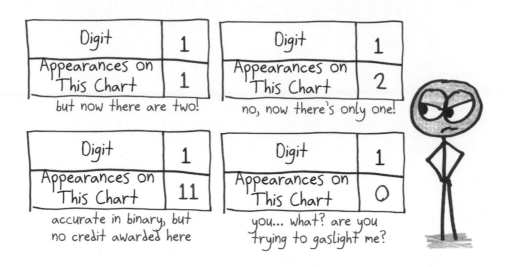

Digit	1
Appearances on This Chart	1

but now there are two!

Digit	1
Appearances on This Chart	2

no, now there's only one!

Digit	1
Appearances on This Chart	11

accurate in binary, but no credit awarded here

Digit	1
Appearances on This Chart	0

you... what? are you trying to gaslight me?

Will adding a 2s column help? Nope. Any attempt to fill the second row will still defeat itself, as surely as uttering the prophecy "No one will ever utter this prophecy."

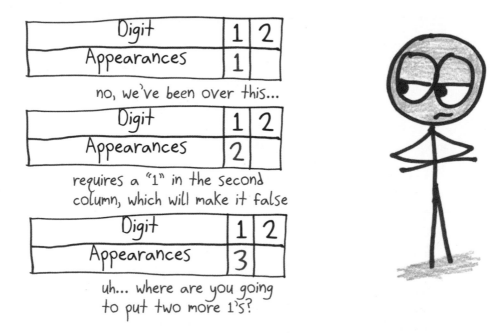

Digit	1	2
Appearances	1	

no, we've been over this...

Digit	1	2
Appearances	2	

requires a "1" in the second column, which will make it false

Digit	1	2
Appearances	3	

uh... where are you going to put two more 1's?

Is such a chart even possible? Yes, but you need a minimum of four columns. Here's one solution; I'll let you discover the other for yourself.[18]

---

18 Longer charts also work. Solutions are in the Bibliography.

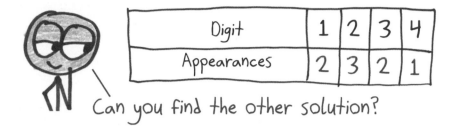

Digit	1	2	3	4
Appearances	2	3	2	1

Can you find the other solution?

Whenever I play Prophecies, my mind falls into recursive loops like these. Will my next move bring about its own downfall? One moment, the numbers feel solid and true; the next, they evaporate like the logic of a dream. The result is a game of mind-bending strategic depth.

WHERE IT COMES FROM

A lovely fellow named Andy Juell.

In 2010, Daniel Solis launched the Thousand-Year Game Design Challenge. The task: Create a simple, profound, and enduring game, the kind that could last for a thousand years. For his entry, Andy Juell submitted Prophecies (which he distilled from his own board game Actual Size May Vary).

Fast-forward to 2020. Not only did Prophecies rank among my play testers' favorite games, but Andy himself turned out to be a witty, humble, and supremely helpful email correspondent, with a dazzling knowledge of strategy games.[19]

As for Prophecies lasting a millennium—hey, why not? You don't need pencils, paper, or even numerals. All it takes is some dirt in which to scratch out tally marks. "Should the very concept of counting elude our descendants," Andy noted, "I humbly submit that the inaccessibility of this game numbers among the least of their innumerable problems."

WHY IT MATTERS

Because playing with paradoxical self-describing numbers helped to usher in the computer age.

The story begins in the late 1800s, when mathematicians undertook a gargantuan project: to trace their whole subject back to its logical foundations. They hoped to reconstitute the vast sprawl of mathematical ideas into a kind of unshakeable tower, each floor resting sturdily upon the floor below, all the way down to irrefutable bedrock—that is, a set of simple assumptions from which all mathematical theorems follow.

The crucial step, of course, was to select the right assumptions.

---

19 He even let me rename his game as Prophecies, despite his having a dozen cleverer and more literary titles at the ready, such as Just Like It Says on the Tin, Count Me In, Scry vs. Scry, and Self-Fulfilling Profits.

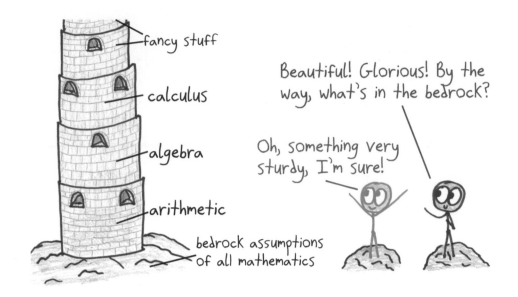

A typical effort might begin with the empty set, a kind of conceptual bag with nothing inside. Make this represent the number 0, and build from there. The set containing the empty set might be called "1." The set containing 1 and the empty set could be called "2." The set containing 2, 1, and the empty set could be called "3." And so on.

Proceed in this fashion, building sets of sets of sets into logical structures of increasing complexity, until you've captured every number, shape, and equation imaginable.

All of mathematics, from a few simple assumptions about sets.

Number We're Defining	Defined as a Set of Numbers	Defined as a Set of Sets
0		{ } (nothing inside)
1	{ 0 }	{ { } }
2	{ 0 1 }	{ { } { { } } }
3	{ 0 1 2 }	{ { } { { } } { { } { { } } } }

Alas, for decades, these efforts floundered and foundered. Paradoxes kept cropping up, forcing mathematicians to alter their foundational assumptions. But the new assumptions only led to new paradoxes, or else left the bedrock too weak to support the mathematical tower.

Finally, in 1930, a logician named Kurt Gödel showed why everyone had been having so much trouble. The whole project was an impossible dream.

His argument goes like this. First, for the bedrock of your mathematical tower, pick any set of assumptions you like (as long as they cover basic arithmetic). Then, Kurt will show how these assumptions create a kind of language, in which statements (such as "0 equals 0") can be encoded as numbers (such as 243,000,000). Finally, Kurt will produce certain statements that refer to their own numbers, saying, more or less, "The statement with this number cannot be proven true."

In a word: Uh-oh.

"This statement cannot be proven true" really cannot be proven true, because such a proof would render it false. Meanwhile, proving it false would show, beyond a shadow of a doubt, that it cannot be proven true. But that proves it *is* true!

Impossible to prove true, impossible to prove false, such statements must lie beyond the reach of proof, in a spooky third category Kurt called "undecidable."

Scholars had hoped to find mathematical bedrock. Kurt set dynamite to this hope. No matter what assumptions you make, there will always be statements you cannot prove, heights to which your tower can never reach.

The most ambitious mathematical project of the century, doomed by self-describing numbers.

After Kurt's bombshell, other mathematicians tried to salvage something from the wreckage. One fellow named Alan Turing envisioned a machine that could help decide which statements are true, which are false, and which are undecidable—a kind of automatic truth determiner.

Or, as we call it today, a computer.

I'm reminded of this lineage every time I play Prophecies. Numbers that describe themselves, numbers that contradict themselves, logical knots, self-referential loops . . . This stuff isn't just for gags and giggles. It's the primordial soup from which the computer age emerged.

Kurt likened his self-referential numbers to the Liar's Paradox. This old brain-tickler says, in its simplest form, *This statement is false*. It can't be true, or we'd have to accept what it says and declare it false. But it can't be false, or we'd have to *reject* what it says and declare it true. The statement thus floats in a gray mist: not true, not false, not living, not dead, not taxed, not tax-exempt. It's a kind of semantic ghost with unfinished business.

Or it was until Kurt came along. That's when this ancient paradox, after millennia of haunting poor logicians, climbed inside a box of transistors and realized its true calling: to haunt us all.

VARIATIONS AND RELATED GAMES

EXOTIC BOARDS: Rather than a grid, you can play on any collection of overlapping regions. Here's an example, with five regions each consisting of 16 cells:

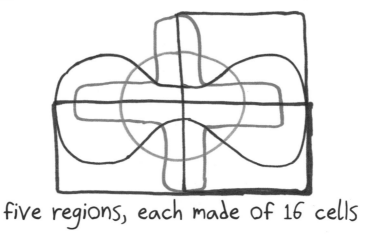

five regions, each made of 16 cells

MULTIPLAYER: The game also works with three or four players. Just use a larger board (e.g., 7 by 7).

X-PROPHECIES: Treat each number as a prediction of the *number of X's* in its row or column (rather than the number of numbers).

SUDOKU BOARD: Possibly the coolest way to play. Use an unsolved sudoku puzzle as your board; each prophecy applies not only to its row and column, but also to the 3-by-3 square where it belongs. Preexisting numbers don't count for either player.

THE BERRY PARADOX: Okay, this isn't a game; I just want to show you one final demon from the Pandora's Box of Self-Referential Numbers.

This paradox begins with the simple observation that larger numbers typically require more letters to describe. For example, I can describe 10 in just three letters ("ten"), but for 1,000 I need eight letters ("ten cubed," which is shorter than the 11-letter description "one thousand"). The librarian G. G. Berry extended this idea by naming "the smallest positive integer not definable in under sixty letters."

It sounds perfectly sensible, as if this number (whatever it is) must exist . . . until you realize that you've just defined it in 57 letters. The definition annihilates itself.

AN ENUMERATION OF NUMBER GAMES

When I began writing this book, it was my honorable intention to highlight five games per section—no more, no less. Why did I abandon this plan? Blame the diverse and multitudinous ranks of nifty little number games. I can't very well leave them out in the rain, can I? Instead, I shall offer seven games in rapid-fire succession, in this zesty sampler platter of a chapter.

MEDIOCRITY

A GAME OF MEDIANS

Mediocrity is a three-player game that was first played by two siblings and a friend on restaurant napkins.[20]

Each player secretly selects a whole number between 0 and 30.[21] Then the choices are revealed, and **whoever chose the median (i.e., middlemost) number wins that number of points**. If two people choose the same number, then the player who picked a different number gets to break the tie and assign the points to whichever other player they wish.

winner! scores 10 points

---

20 Another name for the game is Hruska, after US senator Roman Hruska. He once defended an embattled Supreme Court nominee by saying, "Even if he were mediocre, there are a lot of mediocre judges and people and lawyers. They are entitled to a little representation, aren't they?"
21 This is flexible. Some folks enjoy playing with no limits, meaning you can pick any number at all.

But wait. There's another twist. You play a fixed number of rounds—let's say five—at the end of which **the winner is not the one with the *most* points, but the middlemost number of points**. As co-inventor Douglas Hofstadter put it, that's the only way for "the spirit of the *whole* . . . to be consistent with the spirit of its *parts*."

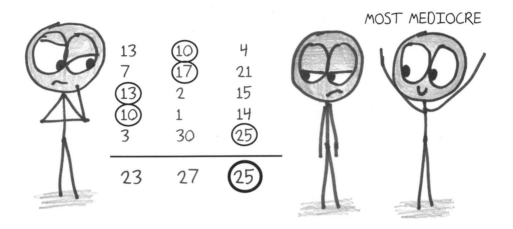

A few final suggestions: (1) I recommend keeping the scores in full view, as these will influence strategy. (2) The game works with any odd number of players;[22] if you have an even number of players, simply insert an invisible player who picks 15 every time. (3) For a truly mind-bending experience, play a tournament of five games, and whoever wins the median number of games wins the tournament.

---

22 Indeed, this is a nice thing about Mediocrity. Too many three-player games are dysfunctional because they incentivize the players in second and third place to team up and attack the player in first. In Mediocrity, no such "ganging up" is possible or desirable.

BLACK HOLE

A GAME OF SUDDEN COLLAPSE

In this two-player, two-color game from Walter Joris, the tension mounts at each move, culminating with an explosive finish. I can't vouch for its cosmological accuracy, but I enjoy its puzzle-like flavor, with victory and defeat hinging on the one space left unfilled.

To begin, draw **a pyramid of 21 circles in six rows**, as shown. Then, take turns writing the number 1 in your color in a circle of your choice. After that, **take turns writing 2, 3, 4, and so on**, going through the numbers in order.

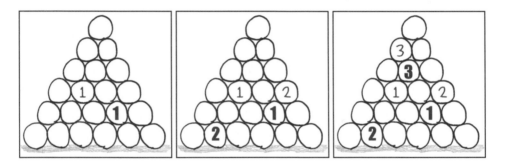

When you have both reached 10, there will be **one circle left blank: the black hole, which immediately destroys all of its neighboring circles**. Whoever has a greater sum of numbers left over—that is, **whoever loses a smaller sum to the black hole**—is the winner.

lost 8   lost 13

Simple as the game is, I've found it hard to develop strategic intuition. You can't "reserve" spots for your high numbers, or your opponent will simply fill them. Somehow, you must leave spots that your opponent doesn't *want* to fill (perhaps because they'd protect your own numbers). But don't leave too many spots like this, or one of them will spell your defeat.

JAM

A GAME OF 15s

This very simple game takes just 31 words to explain. **Take turns claiming numbers from 1 to 9. No number can be claimed twice. The winner is the first player to claim a trio of numbers that add up to 15.**

Jam may remind you of tic-tac-toe. That's not a coincidence. It *is* tic-tac-toe—or, as a 1967 psychology paper put it, "an isomorph of tic-tac-toe." By arranging the numbers 1 to 9 in a magic square (so that every row, column, and diagonal sums to 15), the correspondence becomes clear.

The two games are structural analogues, identical twins wearing different disguises.

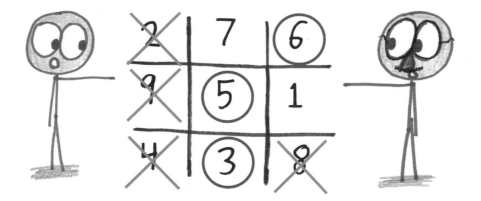

Humans are not, by nature, abstract thinkers. Quite the opposite. We are creatures of the concrete: animals who experience appetites, daydreams, and dizzy spells—sometimes all at once, if the Food Network is on. We think in particulars. That's why isomorphisms matter: They transport us from one set of particulars to another. An isomorphism is an abstract bridge between islands of experience.

While we're at it, here's another game isomorphic to tic-tac-toe. I call it Sir Boss's Barn. To begin, write down the sentence: *In fact, Sir Boss's Barn was built on rot.* Then, take turns circling words. If you claim three words with a letter in common—such as "**in**," "**sir**," and "**built**"—then you win.

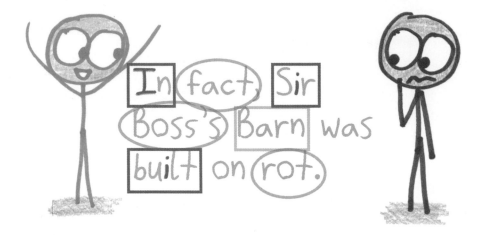

You needn't use this particular sentence.[23] Half the fun is creating your own set of words that are isomorphic to tic-tac-toe.

rot	fact	built		fat	as	pan		tug	us	pun
on	barn	In		if	spit	in		at	gasp	an
Boss	was	Sir		fop	so	not		tip	his	gin

I encourage you to seek others, and hit me up if you find a good one.

---

23 That said, I like that its possible winning letters are A, B, I, N, O, R, S, and T, which can be anagrammed into the purpose of the game: NAB TRIOS. I also like that in the name "Sir Boss's Barn," three sensible words add up to a nonsense title, whereas in "tic-tac-toe," three nonsense words add up to a sensible title.

STARLITAIRE

A GAME OF PRETTY PICTURES

Starlitaire is a little atypical for this book—a one-person pastime with no
rules, no way to win, and no way to lose. Still, I believe the mathematical
artist Vi Hart (who popularized the idea in a viral YouTube video) was quite
right to call it a game. Why? Because a game, at its core, is structured play,
and that's precisely what Starlitaire offers.

Begin with a circle of dots, as many as you like. Then start connecting
them according to a rule of your choosing. For example, you might skip two
dots each time.

Beautiful patterns will emerge as if by magic.

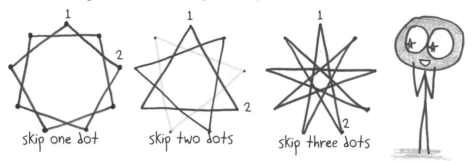

skip one dot   skip two dots   skip three dots

Though it appears geometric, the game's true structure is numerical.
These stars obey the gravitational laws of prime factorization. With 12 dots,
for example, you'll sometimes create distinct, overlapping shapes. But with
13 dots, this never happens.

Why not? Because 13 is prime, and 12 isn't.

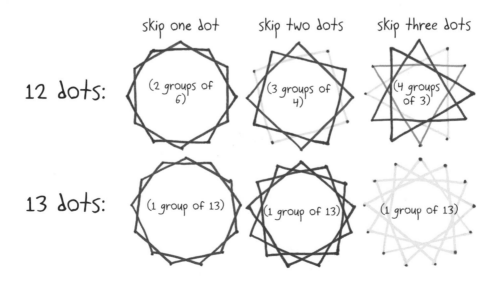

Like mathematics itself, Starlitaire is an inexhaustible playground, a game with no beginning or end. Try it with multiple colors, differently spaced dots, or more complex rules of connection (e.g., skip 1, skip 2, skip 1, skip 2). Just be careful not to lose yourself among the stars.

"Number theorists are like lotus-eaters," warned mathematician Leopold Kronecker. "Having tasted this food they can never give it up."

GRIDLOCK

A GAME OF FILLING ARRAYS

From a collaborative game by Stanford education center YouCubed, I've adapted this sneakily competitive one. You need two standard dice (easily simulated; search the internet for "roll dice") and a 10-by-10 grid for each player.

The goal: Fill as much of your grid as you can before the game comes to an end.

On each turn, roll the two dice. Say you get 4 and 5; you must then **shade in a 4-by-5 rectangle** anywhere on your grid. **If the rectangle will not fit on your board, you lose your turn.** When both players lose their turn one after the other, the game ends, and **whoever has more filled-in squares on their grid is the winner.**

Just one twist: **You may, whenever you wish, choose to play your rectangle on your opponent's board instead of your own**.

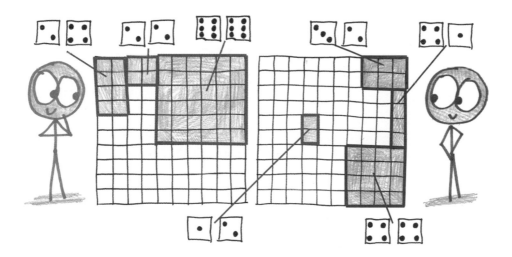

Why play on your opponent's board? Well, as the picture indicates, a small rectangle in the middle of your foe's board can foil their whole Tetris-like plan.

TAX COLLECTOR

A GAME OF CAREFUL DEDUCTIONS

This game has circulated for half a century as an introduction-to-programming exercise. Fifteen years ago, Robert Moniot was already calling it a "golden oldie." It's a solo game—more a collection of puzzles, really—played against a ruthless, automatic opponent: the Tax Collector.

To begin, write down **all of the whole numbers up to some ceiling**, such as 12.

Then, on each turn, **claim a number, and add it to your score. The Tax Collector receives all of the remaining numbers that divide it evenly**.

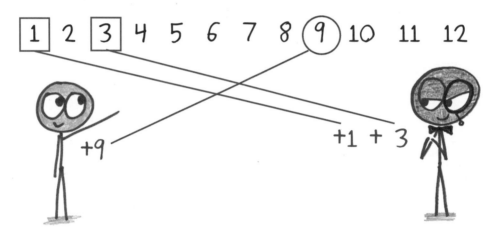

The challenge is that **the Tax Collector must receive something on every turn**. For example, at this point, you cannot pick 5, because its only divisor (namely 1) is already gone.

Keep going until the only remaining numbers have no divisors left, at which point the Tax Collector gets them all.

The goal, of course, is to beat the Tax Collector.

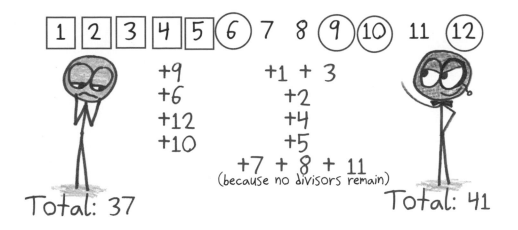

A useful starting point is the "greedy algorithm": On each turn, pick the number that nets you the most points. For example, in the game with a ceiling of 15, the best first pick is 13 (which nets you 12 points).

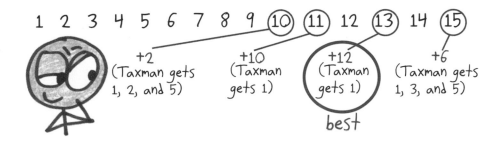

But sometimes this will overlook a superior move. For example, after choosing 13, the greediest remaining choice is 15 (which nets you 7 points). But that'd leave you unable to pick 9, which will later become a gift to the Tax Collector. It's better to pick 9 first, and *then* 15.

For small numbers, you can track your choices by using playing cards (e.g., if your ceiling is 12, then lay out the cards 2 through 10, plus ace = 1, jack = 11, and queen = 12). The game is, even more so than other games, best played while listening to the Beatles' "Taxman."

LOVE AND MARRIAGE

A GAME OF PRECARIOUS PARTNERSHIPS

James Ernest created this matchmaking game at the request of "a middle school teacher who was looking for a good interactive classroom game on the subject of love and marriage." It requires at least 15 players, and is suitable for a nerdy party or partying classroom.

With *n* players (say, 27), begin by creating a **deck of cards numbered 1 through *n* + 10** (in this case, 1 through 37). Also, create a **scoring track**, with rows numbered 100, 95, 90, 85, 80, and so on, all the way down to 5. Leave enough space in each row for two cards.[24]

To begin a round, shuffle the deck, and **give each player a card**. On the word "go," everyone has **three minutes to find a partner and place their paired cards in the highest open spot on** the scoring track.

---

24 With 20 or fewer players, you only need 10 rows, numbered 100, 90, 80, and so on.

Your score is computed by these criteria:

1. **Marrying fast.** You want to marry as soon as you can, because your base score is the label of your row. Early marriers get bases like 90 and 85; late marriers, 15 and 10.

2. **Marrying close.** You want your number to be as close as possible to your partner's, because your base score is then divided by the difference between your two numbers. For example, if 25 marries 28, they divide their base score by 3.

3. **Marrying up.** The lower number in each marriage gets a 5-point bonus. The higher number gets no bonus.

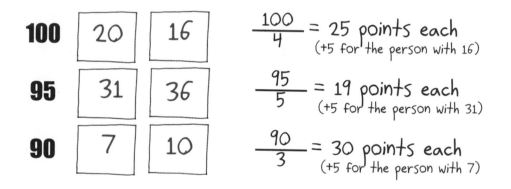

$$\frac{100}{4} = 25 \text{ points each}$$
(+5 for the person with 16)

$$\frac{95}{5} = 19 \text{ points each}$$
(+5 for the person with 31)

$$\frac{90}{3} = 30 \text{ points each}$$
(+5 for the person with 7)

While I don't endorse the semi-adversarial view of marriage, I love the tension. All at once, you want to marry fast, marry close, and marry up. These impulses pull against each other, raising tricky questions. For example, is 1 a good card? On the one hand, you're guaranteed to marry up. But the closest numbers (like 2, 3, and 4) may reject you, hoping to marry up themselves. As in real life, no strategy is "good" or "bad" in isolation; it must be judged in the context of everyone else's choices.

III

COMBINATION GAMES

ACCORDING TO RAPH Koster—game designer, bestselling author, and guy whose name I always misspell as "Ralph"—every game embodies one of four core challenges.

THE FOUR CORE CHALLENGES OF GAMES

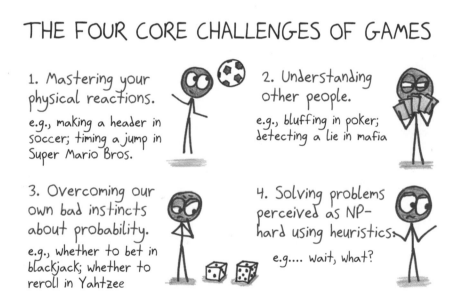

1. Mastering your physical reactions.

e.g., making a header in soccer; timing a jump in Super Mario Bros.

2. Understanding other people.

e.g., bluffing in poker; detecting a lie in mafia

3. Overcoming our own bad instincts about probability.

e.g., whether to bet in blackjack; whether to reroll in Yahtzee

4. Solving problems perceived as NP–hard using heuristics.

e.g.... wait, what?

Raph's quartet may strike you as a little peculiar. He gives three simple truths plus one piece of inscrutable jargon. It's as if he named the most common pets as "dogs, cats, fish, and the pliability of ethical guidelines." Yet Raph is exactly right. There is a spooky connection between combinatorial complexity and our instinctive sense of fun. With eerie, unconscious consistency, we seek a special middle ground: puzzles whose solutions are hard to find, yet easy to appreciate once found.

Explaining this pattern requires a branch of computer science called *complexity theory*. It asks a simple question: When you make a problem bigger, how much harder does it get? Take the famous Traveling Salesman Problem. On the map below, what's the shortest journey you can take through all four cities, ending back where you began?

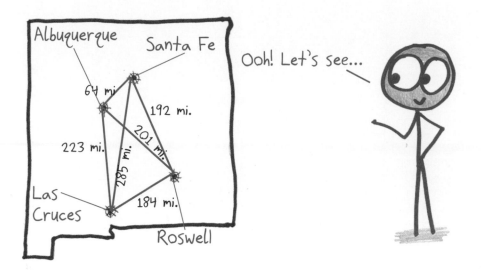

Starting in Albuquerque, you might go to Santa Fe, Roswell, Las Cruces, and home (663 miles); or Santa Fe, Las Cruces, Roswell, and home (734 miles); or Roswell, Santa Fe, Las Cruces, and home (901 miles). Believe it or not, those are all of your options. Although you can write down 21 other loops, each will be the same as one of these three, except from a different starting point, or running in reverse order. So the shortest is going from Albuquerque to Santa Fe to Roswell to Las Cruces and then back to Albuquerque. Problem solved.

Now, what if I add three more cities? In this version, there are 360 loops to consider. It's unwieldy by hand, though a computer can make quick work of them.

What if we go further? Make it, say, all 37 cities in New Mexico?

There are now almost 200 duodecillion paths, or 2×10^{41}. No way you can explore them all; even a Los Alamos supercomputer would collapse like a camel's back under 10^{41} straws. This is a brutal and fundamental pattern in math: the *combinatorial explosion*. Add a few extra items, get zillions of new combinations.

To solve a problem by checking every possible combination is known as *brute force*, and thanks to the combinatorial explosion, it is slower than a sloth in a molasses bath. A computer that can brute-force the 10-city problem in a single second will barely manage the 20-city problem in a hundred centuries. A complexity theorist would say the brute-force algorithm "runs in exponential time." This means "extremely slow," and is a useful bit of jargon if your computer scientist friends are ever running late for lunch.

Luckily, we have better options than brute force. Sometimes much better. The quickest-to-solve problems, such as multiplying numbers or sorting a bunch of objects by size, belong to a category called P, for "polynomial time."

These problems can be solved in a reasonable amount of time.

P

multiplication

sorting

some matrix operations

A larger category is NP (for "nondeterministic polynomial time"). With these problems, it's quick to verify a solution, but not necessarily to find the solution in the first place. Many of our favorite games and puzzles fall into this category. Sudoku is a good example. The solution is elusive, requiring clever feats of deduction, yet it's easy to verify, requiring only that you check every row, column, and 3-by-3 box for the numbers 1 to 9.

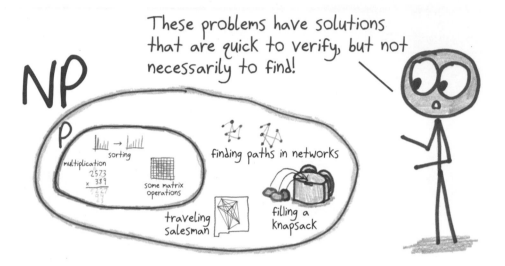

I'm obligated to mention here a deep unsolved problem in math, worth $1 million to whoever cracks it: P vs. NP. Are P and NP truly different, or secretly the same? Most experts think they're different; the vexing challenges of NP feel much harder than the easy pickings of P. But no one has proven it. There's a slim chance that some brilliant undiscovered algorithm will solve all of those NP problems in one fell swoop.

In the meantime, there remains a mysterious correspondence between complexity and fun. Somehow, just as our brains are drawn to sugar, gossip, and glowing screens, they're drawn to NP-hardness.

Take the tile-sliding Fifteen Puzzle. Its debut in 1880 launched a worldwide craze. "Thousands of men," reported the *New York Times*, "who but lately were honest and industrious have yielded to its fatal fascinations, and neglecting their business and their families, spend their whole time over the demoralizing box."

It stood as the most popular puzzle of all time for a century, until 1980, and the advent of another combinatorial toy: the Rubik's Cube. Before long, Rubik's-themed books occupied spots 1, 2, and 5 on the *New York Times* bestseller list. The cube soon became the bestselling toy in human history, a title it retains nearly 400 million units later. ABC even aired a Saturday-morning cartoon titled *Rubik, the Amazing Cube.*[1]

---

1   It starred a sentient alien shaped like a Rubik's Cube. When in solved position, it could fly, speak, and emit magical space beams. When scrambled, it could mutter only soft gibberish and the occasional cry of "help." Try not to picture that next time you're struggling to solve a cube.

As an algebra teacher, I know how much the masses adore abstract math. It is their third favorite pastime, behind only parallel parking and being taxidermied.

What, then, attracts otherwise sane humans to such vexing puzzles?

I credit the siren song of NP-hardness: the tantalizing complexity of solutions that are slow to find, yet quick to verify. On some unconscious level, our gameplaying instinct is a kind of mathematical instinct, a nose for deep combinatorial waters.

Every game, on one level or another, is a game of combinations.

Game	What You're Combining	The ___, the Better
Poker	Cards	More Aces
Scrabble	Letters	More Plausibly a Word
Jenga	Blocks	More Stable
Twister	Limbs	Fewer Displaced Internal Organs
Dominos	Dominos	More Dominos
Cards Against Humanity	Offensive Tropes	Less You Play This Game

In the coming pages, I'll invite you to join in the combinatorial fun. Should your opponent defeat you with a move you never considered, yet whose brilliance feels obvious in hindsight, take heart: You just experienced NP-hardness in action.

SIM

HOW TO GIVE THE WORLD A HEADACHE WITH JUST SIX DOTS

Sim takes its name from Gustavus Simmons, the Albuquerque mathematician who first analyzed it. Also from a word whose relevance will soon become obvious: "simple."

Still, the game's deeper origins lie in the wild and expansive mathematics of Ramsey theory.

Ramsey theory is named for Frank Ramsey, who was born in 1903 and died in 1930. In his few, fruitful years, he advanced the fields of economic theory, probability theory, and logical paradox. He even befriended philosopher Ludwig Wittgenstein, whom historians describe as "unbearable." Yet for all those accomplishments, the theory to which Ramsey lends his name deals with a seeming triviality: games of multicolor connect the dots. *If I want to guarantee the presence of a certain shape,* Ramsey theory asks, *exactly how many dots do I need?*

Sound silly? Guilty as charged. Sound simple? Careful: Don't let the name fool you.

HOW TO PLAY

What do you need? Two players, each with their own distinct color pen, and six dots drawn as below.

What's the goal? Force your opponent to create a triangle in their color before you're forced to create one in yours.

What are the rules?

1. Take turns **connecting any two dots using your colored pen**.

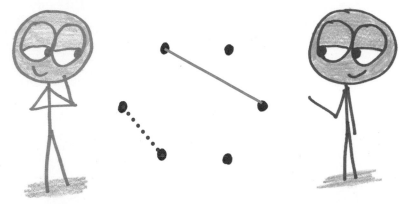

A few of the lines will crisscross. That's okay. Note that each dot will allow at most five connections, i.e., one to every other dot.

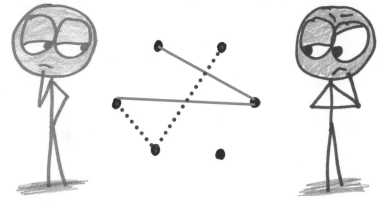

2. If your opponent **creates a triangle entirely in their color**, tap the three dots and spell S-I-M. Congratulations: **You've won!**

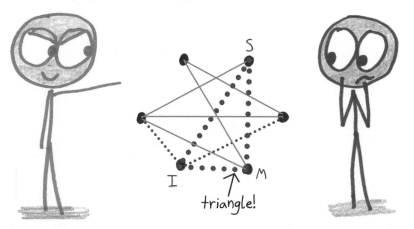

3. If you complete a triangle in your color, don't despair. If your opponent fails to notice it, and moves instead, **you can "steal" the victory by pointing out the triangle yourself**.

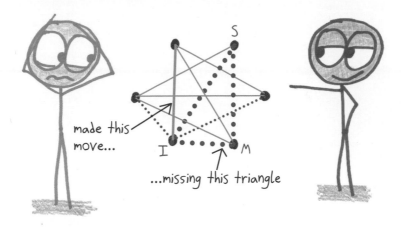

4. However, if you say "S-I-M" with no such one-color triangle present, **you immediately lose the game**.

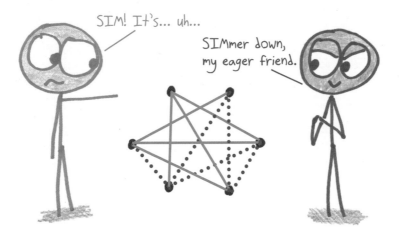

TASTING NOTES

Sim is syrupy sweet when it first hits the tongue. There are plenty of places to move.

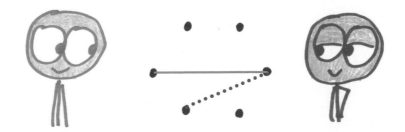

Soon, though, it develops the brooding complexity of an aged wine. In particular, your board may become cluttered with **"distractor" or "decoy" triangles**. They *look* like triangles—I guess because they are—but they don't count for Sim, because in this game, triangles must connect three of the original dots.

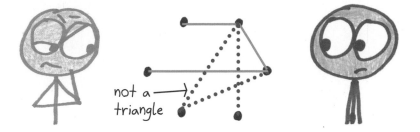

Those six dots allow for 20 possible triangles. That's a lot of traps to evade. You'll stare at the board with furious intensity, as if inspecting the world's smallest crime scene, until . . . aha!

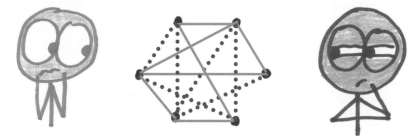

See it? Blue has created a triangle. Here, I'll highlight it:

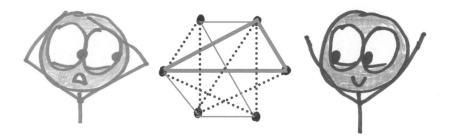

See, I warned you. Six dots can get pretty tricky.

WHERE IT COMES FROM

Ramsey theory. The thrust of the subject is that simple settings can give rise to daunting complexities. For example, here is a brief excerpt from the first published solution for Sim:

Rule 2. For any move other than the second when there is at least one neutral edge, consider only these neutral edges and apply the following rules in a hierarchy...:
(1) Ruin a minimum number of valid safe moves.
(2) Create a minimum number of losers (valid & hypothetical).
(3) Ruin a minimum number of hypothetically safe moves.
(4) Complete a maximum number of mixed triangles.
(5) Create a maximum number of partial mixed triangles.
(6) Create a minimum number of valid losers.
Then color any one of the edges satisfying the above rules.

Uh... what?

This paper, in the journal *Mathematics Magazine*, spells out a guaranteed winning strategy (for player 2, it turns out). Yet quick and contained as the game may be, the strategy is too intricate to memorize.

That's okay. Ramsey theory isn't about how to win. It's about how to design a game where ties are impossible.

It's about guaranteeing that *somebody* wins.

In Sim, specifically, we want someone to draw a triangle all in one color. How many dots does that require? Six dots suffice (for reasons we'll see later), but five do not. Sim on a pentagon might end in a tie.

The game is over... and yet...   ...no one-color triangle!

Sim isn't the only game on the shelves of Uncle Ramsey's Game Shop. To create a new game, we can simply change the aim. What if the losing shape is not *three* dots all connected, but *four* dots all connected—a kind of crossed square, in red or blue? How many dots to guarantee this shape?

The magic number, if you want to know, is 18 dots. And if you don't want to know, my apologies, but it's still 18.

If Sim is a headache, then this crossed-square variant is a hatchet to the skull. It involves 153 moves, 3,060 possible four-dot combinations, and so many crisscrosses that it is, for all practical purposes, unplayable.

Yet Ramsey can crank up the dial even further. What if, instead of four dots all connected by a single color, we're trying to avoid *five*? In other words, what if the losing shape is an outlined star? How many dots to guarantee a winner in *this* game?

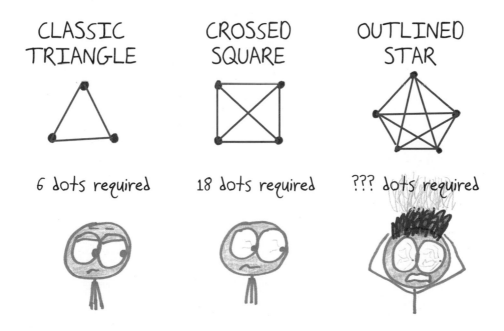

Trick question. Analyzing this game is so hard that, after half a century of work, nobody knows the answer. We know 42 dots aren't enough. We know 48 dots definitely are. Beyond that, we're stumped.

So easy to ask, yet so hard to answer, it's a question you can't help stumbling across.

"I believe Ramsey theory gets invented on most planets that develop space-faring civilizations," muses the mathematician Jim Propp. "Indeed, one of the animating impulses behind Ramsey theory can be experienced when one looks up at the night sky, observes geometric patterns in the stars, and wonders 'How many of these seemingly meaningful patterns actually become inevitable once you have enough stars in your sky?'"

The mathematician Paul Erdős once imagined those same aliens visiting Earth and giving us a year to solve the five-dot outlined star problem, pledging to obliterate us if we fail. "We could marshal the world's best minds and fastest computers," Erdős said, "and within a year we could probably calculate the value."

What about the next step: the Ramsey game where you seek to avoid *six* dots, all connected by your color? If the aliens demanded it, could we calculate the necessary number? Erdős was not optimistic. "We would have no choice," he said, "but to launch a preemptive attack."

That's Ramsey theory for you. It's a land where frivolous games lead to bottomless wormholes, where just six dots can give the whole cosmos a headache.

WHY IT MATTERS

Because we are the dots.

In the 1950s, the Hungarian sociologist Sándor Szalai was watching groups of children.[2] He noticed a strange pattern: Among a class of 20 or so, he could always find a group of four, all of whom were friends, or else a group of four, *none* of whom were friends.

What explained these clusters? Why these mysterious quartets of friendship and alienation? Did it depend on the children's ages? On the specific culture of the school? On the presence of four-square courts?

Then it struck Szalai. Maybe this isn't a sociological fact at all. Maybe it's a mathematical one.

Without knowing it, Szalai was playing the crossed-square version of Sim. Instead of dots: children. Instead of blue: "friends." And instead of orange: "not friends."

"Social network" is a literal term. Everyone you know is a dot, also called a node or a vertex. We share links of various kinds. Blue for acquaintances; orange for strangers. Blue for having shaken hands at some point; orange for not. Blue for speaking a language in common; orange for needing to communicate via gestures.

The game of Sim, in this light, becomes a study in six-person groups. It promises that every group of six will contain either (a) three mutual friends, or (b) three mutual non-friends.

To see why, pick one person from the six, and call this person Dorothy (or Dottie for short). Then pencil in Dottie's connections to the other people: blue for friends, orange for not.

---

2   This is a better research method than the reverse, in which children watch groups of sociologists.

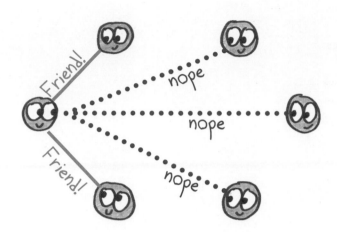

With five connections in all, at least three will necessarily be the same color. (I'll go with orange, but it works either way.) If any of these three people share an orange connection, then we have our group: Dottie plus that pair. None of them are friends.

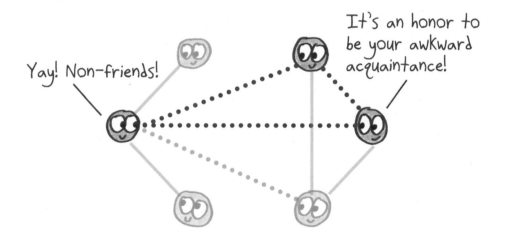

Meanwhile, if *none* of these three share an orange connection, then they must all share blue connections. Which means we have our group: those three mutual friends.

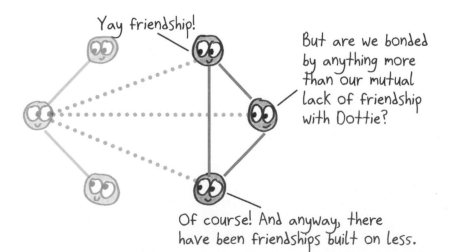

Anthropologist Robin Dunbar has suggested that the human brain evolved to manage only a finite number of relationships: 150, give or take. That's the rough size of a hunter-gatherer tribe. This figure may sound measly to the modern city-dweller. I've got twice that many contacts on LinkedIn, and I can't stand LinkedIn. How did our ancestors survive their claustrophobic world? Was life a constant agony of loneliness, boredom, and everyone dating their friends' exes?

Well, consider: In a tribe of 150, there are more than 11,000 personal relationships, more than 500,000 possible trios, and more than 20 million possible quartets. The numbers only climb from there. The variety of potential cliques, schisms, and alliances is dumbfounding.

A social world of 150 is not simple. It is not even complex.

It is unfathomable.

By the way, you know the number that Erdős said we'd never find—the number of dots needed to guarantee an all-orange or all-blue group of six? It is known to be somewhere between 102 and 165 dots. That's pretty much the size of those hunter-gatherer tribes.

We humans evolved to live in networks of a dozen dozens: little tribes so vast that mathematics will never fully grasp them.

VARIATIONS AND RELATED GAMES

WHIM SIM (FOR THREE PLAYERS): Want to add a third player? You could introduce a third color, but you'd need 17 dots to guarantee an eventual triangle, which makes for a very messy board. An easier option: Stick to two colors, and on your turn, you can pick either of the pens to use, according to your own whims.

JIM SIM (FOR TWO PLAYERS): Mathematically, it doesn't matter how you arrange the dots. It's a game of connections. Shifting the seating arrangement doesn't alter who's friends with whom.

Visually, though, the arrangements differ. Some six-dot formations create bigger headaches than others.

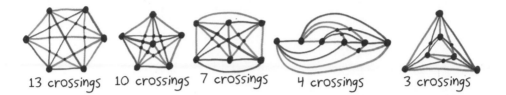

13 crossings   10 crossings   7 crossings   4 crossings   3 crossings

To me, the worst headaches come from "decoy triangles": those created by crossing lines, rather than by connections of the original six dots. In this sense, the standard hexagonal board is a nasty offender, with 13 crossings and 90 decoy triangles.

That's why my father, Jim Orlin, suggested a new formation: the triangle-within-a-triangle board you see above on the far right. It minimizes the number of crossings (there are just three) which thus minimizes the number of decoy triangles (there are 12). I dub this variant Jim Sim in his honor.

I'm not saying Jim Sim creates *no* headaches. Just more manageable ones.

Lim Sim[3] (for a Crowd): In an email, Glen Lim informed me that math camps have been playing a variant of Sim for years. You split into three or four groups, then draw a large number of dots (perhaps 15 to 20) on a whiteboard or sheet of chart paper. Each team, on its turn, goes up to the board and connects a pair of dots using their assigned color. A time limit (say 15 seconds per turn) adds suspense, especially as the game advances, and teams shout frantically about which dots to connect. You can play either fewest triangles wins or most triangles wins.

---

3   Not to be confused with mathematician Courtney Gibbons's suggestion of what to do if you run out of paper: Draw the game board on your arm, creating a "Sim limb."

TEEKO

A NEW GAME FROM OLD PARTS

"The relative merits of Chess and Checkers," wrote the late magician John Scarne, "have been discussed by the millions of followers of these games over the centuries." I found this surprising to read—it sounded a bit like debating the vocal merits of Ariana Grande vs. the guy from Smash Mouth—but not half as surprising as what came next. "With the arrival of Teeko," Scarne declared, "this debate has become three-sided."

The arrival of what now?

Teeko is a simple board game, a blend of several classics, invented by—guess who?—John Scarne. So deeply did he believe in the game that he named his own son Teeko. "If my dad had invented the game of checkers," he explained, "I would be proud to be named Checkers." It's a sentiment I hope to hear echoed one day by my daughter, Bad Drawings Orlin.

"Teeko will undoubtedly achieve ranking," Scarne concluded, "as one of the great games of all time." Well . . . I'll let you and BD be the judge of that.

HOW TO PLAY

What do you need? Two players and a 5-by-5 grid (you can use part of a chessboard, or just draw the grid on a sheet of paper). Then, four tokens of one kind, and four tokens of another. Options include black vs. red checkers, black vs. white pawns, pennies vs. nickels, pennies vs. penne pasta, and rubies vs. emeralds.[4]

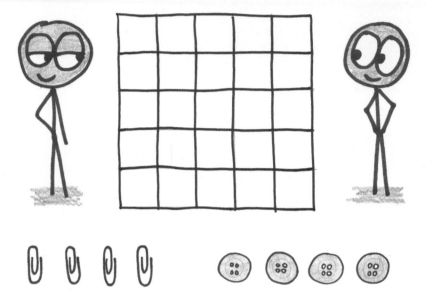

What's the goal? Create a four-in-a-row, or alternatively, a square.

What are the rules?

1. **Take turns placing a token in an empty spot, until all tokens have been placed.** Barring beginners' mistakes, this phase of the game won't produce a winner.

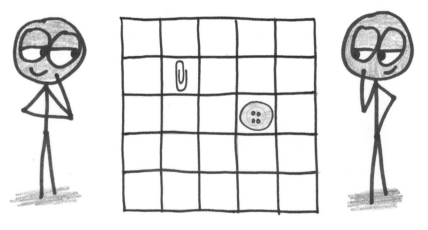

---

4   One of my play testers used chunks of staples. She does not recommend it.

2. Now that all eight tokens are in play, **take turns moving any one of your tokens one step in any direction.** You can only move to an empty space. There's no capturing, and no passing a turn.

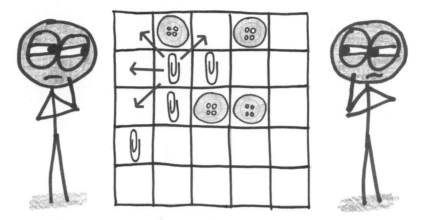

3. The winner is whoever creates either (a) **four in a row . . .**

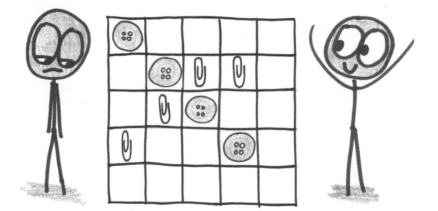

. . . or (b) **the four corners of a square**. The square can be any size (from 2 by 2 to 5 by 5), as long as its edges are vertical and horizontal. No "tilted" squares are allowed.

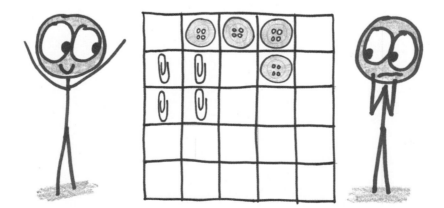

TASTING NOTES

Victories in Teeko come suddenly. There is no distinctive endgame, no gradual capture of enemy pieces. It's more like a jack-in-the-box. Turn the crank, nothing happens. Turn the crank, nothing happens. Turn the—*pop!* Victory!

You may want to implement a "no surprises" rule. That is, if you achieve a position that threatens victory on the next move, you must warn your opponent by calling "check" (like in chess). This prevents the game from ending on a silly oversight.

Either way, Teeko is sweet, snackable, and intermittently spicy. It's like a bowl of honey-roasted nuts where one in thirty has an invisible coating of wasabi.

WHERE IT COMES FROM

John Scarne first published Teeko in 1937, and kept tweaking it into the 1960s. It was a pet project, a lifelong dream, and a desperate bid for what Scarne craved most: legitimacy.

Scarne had come up through the gambling underworld. Never mind that he became the United States' foremost card magician, a friend of Harry Houdini's, and a fixture of TV variety shows. His seedy past followed him like a shadow (or so he felt). Hence his turn to Teeko, a sweet, wholesome, fun-for-the-whole-family sort of game. No deception, and no mafiosos.

In the 1950s, as Teeko attracted celebrity players such as Humphrey Bogart and Marilyn Monroe, it seemed poised to fulfill Scarne's wild ambitions. But its heyday passed. It's an obscurity today, known only to a few expert gamers (and now to you). Not exactly the chess-slayer that Scarne hoped it would become.

WHY IT MATTERS

Because a combination without context is meaningless.

The old adage tells us that a few monkeys with typewriters would, given eternity, reproduce the complete works of Shakespeare. The idea is that every piece of text, be it a classic like *Romeo and Juliet* or a dud like *Timon of Athens*, is in the end a combination of letters.

There's just one snag: Monkeys don't speak English. To them, "to be or not to be" is gibberish, indistinguishable from "uuyfneuzqs" or "lxqjy ubl" or "vote third party for president." For a combination of letters to gain beauty and power, it needs context—that is, a set of relations to *other* combinations of letters. Lacking that, it's like English literature to a monkey, or monkey literature to an Englishman. A jumble in a sea of jumbles.

So, back to Teeko.

By Scarne's own description, Teeko is a Franken-game, stitched together from older classics. "I occasionally describe Teeko as being a combination of four games: Tic-tac-toe, Checkers, Chess and Bingo," he wrote. "The opening moves remind me of Tic-tac-toe; the diagonal moves, of Checkers; the forward, back and side moves, of Chess; and the winning positions are reminiscent of Bingo."

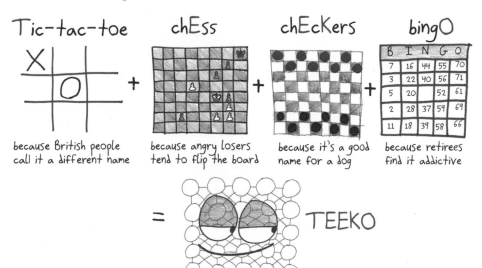

This origin story reminds me of a Jonathan Coulton lyric, sung by a mad scientist trying to woo a romantic partner:

> *I made this half-pony, half-monkey monster*
>   *to please you.*
> *But I get the feeling that you don't like it—*
> *what's with all the screaming?*
> *You like monkeys.*
> *You like ponies.*
> *Maybe you don't like monsters so much . . .*
> *Maybe I used too many monkeys . . .*
> *Isn't it enough to know that I ruined a pony*
> *making a gift for you?*

To create via combination, you need to understand the elements you're combining. I don't think Scarne did. When Omar Khayyam wrote, "We are in truth but pieces on this chess board of life," was he thinking "because we make forward, back and side moves"? Or consider Scarne's efforts to stir up a checkers vs. chess rivalry. According to him, one key pro-chess argument is that chess has more pieces. That's it. That's the argument. More pieces, better game. A monkey with a typewriter could craft a stronger case on a two-hour deadline, even if glued to a pony.

Scarne often boasted of Teeko's "1,081,575 different playing positions." Strangely, that's a lowball figure: The actual number is 70 times larger,[5] making Scarne guilty, for perhaps the only time in his life, of understatement. But even using the correct number, Teeko's complexity pales in comparison to its rival games.

Game	Positions	If each position were an atom, the state space would be the size of...
Teeko	75 million (i.e., 75,000,000)	a bacterium
Checkers	500 quintillion (i.e., 500,000,000,000,000,000,000)	a housefly
Chess	10 tredecillion (i.e., 10,000,000,000,000,000,000,000, 000,000,000,000,000,000,000)	One of the Great Lakes
Go	50 duovigintillion googol (i.e., 500,000,000,000,000,000,000, 000,000,000,000,000,000,000,000,000, 000,000,000,000,000,000,000,000,000, 000,000,000,000,000,000,000,000,000, 000,000,000,000,000,000,000,000,000, 000,000,000,000,000,000,000,000,000, 000,000,000,000,000)	picture the visible universe filled with sand, and then... yeah, I give up, it's too big

More important than the sheer number of combinations is how they relate. What context do they create for one another? In chess or checkers, they create a narrative arc. From the standard starting position, each game builds to a zenith of complexity; after that, the board empties out, culminating in a tense and puzzle-like endgame. One glance tells you exactly how much of the story has elapsed.

---

5   Scarne calculated the number of ways to arrange eight identical tokens on a 5-by-5 board, rather than four of one kind plus four of another.

Not so for Teeko. Look at a game underway, and you can't tell whether it's the 5th move, the 50th, or the 500,000th. "Chess and checkers have a sense of progression through different phases," writes Timothy Johnson, who helped me play-test Teeko. "This game does not; it's two players studying endless variations of the same positions, until finally one person reaches a breakthrough."

In Teeko, the combinations lack history. They have no context.

Credit where credit's due: Scarne gazed into the vast, daunting realm of all possible games and plucked out an elegant little combination of rules. That's creative work. But then he proclaimed his game the greatest thing since pizza and weekends. That leaves him in the sorry state of a monkey on a typewriter, to whom *Hamlet* is just another combination of letters, a jumble in a sea of jumbles, a work of monkey literature.

VARIATIONS AND RELATED GAMES

ACHI: This traditional Ghanaian game sits halfway between Teeko and tic-tac-toe. You play on a 3-by-3 grid, and take turns placing four pieces each. Then, once all the pieces are placed, take turns moving, as in Teeko. The winner is whoever makes three in a row.

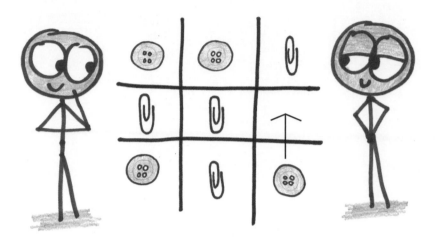

ALL QUEENS CHESS: This game was designed by Elliot Rudell and published by the Happy Puzzle Company. It resembles Teeko, with a few differences: (1) Each player has six pieces, instead of four. (2) They move like queens in chess—that is, as far as you want in any direction. (3) You win by getting four in a row; no victory by square. (4) The pieces begin as pictured below.

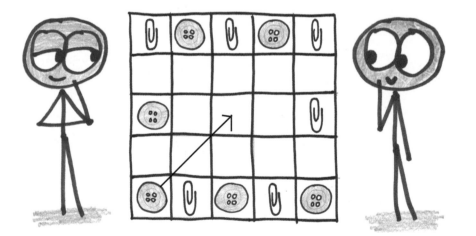

TEEKO CLASSIC: In this chapter, I've presented a variant that Scarne called Teeko Advanced. The original differs in one rule: The only winning squares are the 2-by-2 variety. No larger ones are allowed. The practical difference is small (since it's hard to win via large square anyway), but the theoretical gap is substantial. In 1998, Guy Steele determined via computer analysis that if both sides play perfectly, Teeko Advanced is a first-player win, while Teeko Classic is a draw.

NEIGHBORS

A NUMERICAL MAKE-YOUR-OWN SUNDAE BAR

I have played games in the classroom, and I have played games at parties, and it is a truth universally acknowledged that classroom games rarely work at parties.

Luckily, I know of a glorious exception. It's got Boggle energy, Friday-before-spring-break energy, thinking-as-hard-as-you-want-but-no-harder energy. Play it with your classmates, your friends—heck, play it with your neighbors.

HOW TO PLAY

What do you need? As many players as you like. I've tried 30, but I'm sure you could go higher. (At the other extreme, you can play solo, and try to beat your previous high score.) Each player needs a **5-by-5 grid** and a pen. You also need **one 10-sided die** for the group, which is easy to simulate online (search the internet for "roll dice"). Or you can use a deck of cards (see the VARIATIONS AND RELATED GAMES section).

What's the goal? Place identical numbers in neighboring cells.

What are the rules?

1. The die is rolled, and the result is announced. **Every player writes this
    number in an empty spot somewhere on their personal grid.**

2. This **repeats 25 times, until the grid is full**. You must write down
    each number as it is announced; no saving it or leaving blanks.

3. Now, the scoring. **You score points whenever like numbers (e.g.,
    4-4, or 7-7-7) appear as neighbors within a row or column.** If that
    happens, add their sum to your score.

It's easiest to **score row by row, and then column by column**.

A single **number may count twice**: once in its row, and once in its column.

4. **The highest total score wins.**

TASTING NOTES

It takes a few moves for Neighbors to get humming, but once it does, it keeps a great tempo. It's as if the numbers are vying to expand their territory, and your job is to adjudicate rival claims. Which possibilities will you keep open? Which ones will you foreclose?

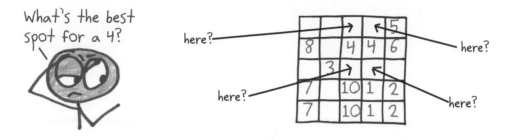

The fact that each number can score twice (once in the row, and once in the column) creates strategic pressure. Some arrangements (e.g., four 3s in a square) are far more valuable than others (e.g., those same 3s in a row).

What never ceases to surprise me is the variety of results. In dozens of games, I've never experienced a tie. As with a make-your-own sundae bar, each person draws on the same few ingredients, yet creates a customized, personal dish.

WHERE IT COMES FROM

Neighbors has rattled around Minnesota for years, whispered telephone-style between math teachers. I learned it in 2019 (under the name Five by Five) from Matt Donald, who learned it in 2015 from Sara VanDerWerf, who learned it in 1991 from Jane Kostik, who learned it in 1987 from . . . a workshop, maybe? Unclear. The Gopher State keeps its secrets.

Anyway, Neighbors was clearly inspired by a classic word game, variously known as Think of a Letter, Crosswords, and Wordsworth:

1. Every player begins with a blank 5-by-5 grid. Players **take turns calling out a letter of their choice**.

2. You write every called letter on your board, in any blank spot you like. Your goal is to **form words across the rows or down the columns**.

3. Three- and four-letter words score their number of letters, while **five-letter words score double** (i.e., 10 points). You may have **at most one word per row and one per column**.

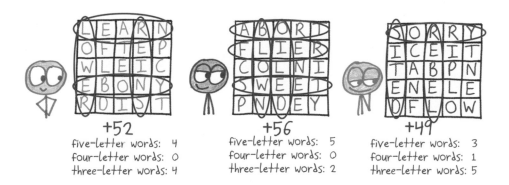

+52
five-letter words: 4
four-letter words: 0
three-letter words: 4

+56
five-letter words: 5
four-letter words: 0
three-letter words: 2

+49
five-letter words: 3
four-letter words: 1
three-letter words: 5

It's not hard to imagine how Wordsworth birthed Neighbors. Sometime in the 1970s or 1980s, a Minnesota math teacher finished a round of this nifty word game, stroked their chin, and imagined replacing the letters with numbers.

Frankly, I'm shocked that it works so well. Which brings me to . . .

WHY IT MATTERS

Because simple combinations create startling variety.

Case in point: the Rocks Game. "Everyone gets some rocks," explains Misha Glouberman in his book (with Sheila Heti) *The Chairs Are Where the People Go*. "You take turns either placing a rock on the floor somewhere or moving a rock that's already there . . . [T]ry not to communicate in any way other than the movement and placement of these rocks. No talking or facial expressions or pointing."

That's it. No victory. No defeat. It sounds like the kind of pointless chore that Greek gods might assign to torture an uppity mortal.

And yet . . .

"Aesthetics emerge very quickly," Misha writes. "So when someone places a rock down, you might think, *Aha! Wasn't that a brilliant thing to do with a rock!* Or, *That ruins everything.* Or, *That was kind of boring.*"

I am no visionary. I would never have invented the Rocks Game, and I could play a thousand rounds of Wordsworth without hitting upon the idea for Neighbors. In fact, I'd have scoffed at the suggestion. Wordsworth uses 26 symbols that can form, depending on your dictionary, over 20,000 different point-scoring combinations. You want to chuck all that in favor of 10 symbols that score by simple repetition? Doesn't that sound dull?

Well, it isn't. A typical round of Neighbors allows for quintillions of arrangements. You could play for eons before two players would produce the same board. As often happens in games, the sharp constraints don't quash strategy. They necessitate it.

For example, if you're placing a 1 on this almost-finished board, which spot is better?

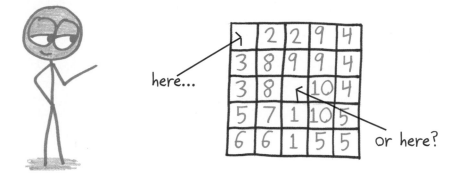

here...

or here?

The center will score you a point. The corner won't. Case closed, right?

Not so fast. If the final number is an 8, 9, or 10, that center square could earn you 16, 18, or even 20 points. The corner is not nearly as promising. If you played this board a million times over, that center square will score an average of 5.5 points; the corner square, a mere 0.5. Thus, it's better to place the 1 in the corner. Never mind the lost point; the real loss would be selling prime real estate to such a low bidder.

At first glance, there's little in common between the structured, probabilistic game of Neighbors and the artsy, improvisational game of Rocks. Yet they share a certain core, a combinatorial engine, that weaves beautiful patterns from simple elements. From the humblest of ingredients, combinatorics can create a vast and varied feast, a veritable ice cream sundae for the mind.

VARIATIONS AND RELATED GAMES

OLD-SCHOOL NEIGHBORS: My pal Matt Donald taught me to play Neighbors using a 10-sided die. But the original rules call for a different randomization method: namely, a deck of cards. You remove the jacks, queens, and kings, and then count the aces as 1s. Put used cards in a discard pile as you go.

The game's feel and flow remain the same, but the underlying probabilities don't. With dice, every number has a 10% chance of appearing on each roll, no matter how many times it has appeared already. With cards, each appearance makes further repetitions less likely.

OPEN BOARDS: Usually, players keep their boards secret until the end of the game. But with only a few players, displaying them openly (and keeping score as you go) can heighten the drama.

WORDSWORTH: This word game is the ancestor and progenitor of Neighbors. It is described under WHERE IT COMES FROM.

CORNERS

A GAME OF PATTERNS HIDDEN IN PLAIN SIGHT

Before we get to the game, here's a warm-up puzzle. In the field of 49 dots below, how many different sizes of square can you find?

Done? Well, I suspect that you missed a bunch. Don't take it too hard. Seeing well is an expert-level skill. That's what makes our next game so tricky to master. Play tester Scott Mittman said it best: "Much has been made about the human capacity to see intricate patterns where none exist (e.g., constellations, clouds, inkblots, economic data) . . . But playing this game brought to light a seemingly contradictory disposition: the inability to see simple patterns where they do exist."

HOW TO PLAY

What do you need? Two players each with their own color pen, and a square grid. I like 7 by 7, but other sizes (such as 8 by 8) work, too.

What's the goal? Create squares, scoring 1 point per corner.

What are the rules?

1. Take turns **marking spaces with an empty dot of your color**.[6] These empty dots are not worth any points—not yet, anyway.

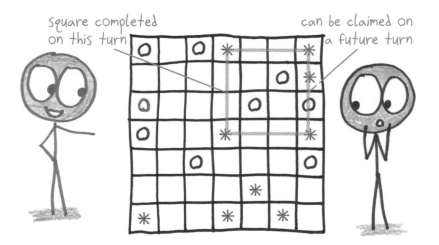

2. If you manage to place **four dots in your color to form the corners of a square**, then congratulations! **On a future turn, you may "claim" this square**.

    square completed on this turn

    can be claimed on a future turn

3. **Claiming a square consumes a turn**, and involves two steps: (a) **Fully shade in the corner dots**, which are now worth 1 point each, and (b) **place empty dots in any unoccupied spaces** in the square.

---

6 My colorblind test-readers flagged this game as tricky to follow. To help clarify, I'll draw Player 1 (green, on the left) using unshaded and shaded asterisks, instead of empty and filled dots.

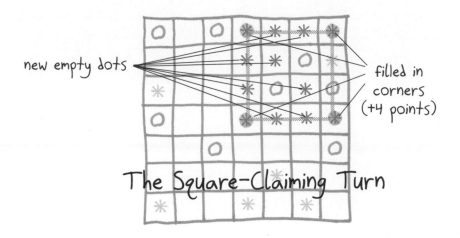

new empty dots

filled in
corners
(+4 points)

The Square-Claiming Turn

4. Note that **squares can occur along 45° diagonals**, too, so that they look
    like diamonds. You may claim a square even if you have already shaded
    some of its corners, and even if there are no unoccupied interior spots.

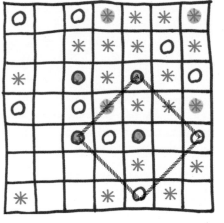

5. **When the board is filled, each player gets one final chance to
    claim a square**. After that, any unclaimed squares remain unclaimed.
    The player with **more shaded dots (i.e., more corners) wins**.

10 points

18 points

TASTING NOTES

On one level, Corners is a game of wide expanses and open fields, a game of *large squares*. Early on, claiming a large square can secure an easy victory by giving you loads of bonus dots in the interior. It's like catching the golden snitch after two minutes.

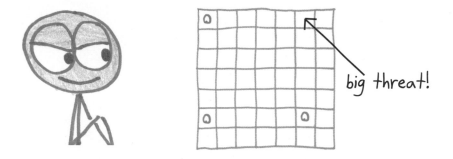

Yet Corners is also a game of tight quarters and quick movements, a game of *small squares*. That's because, unlike with their big siblings, you can threaten multiple small squares simultaneously. That makes them strategically appealing.

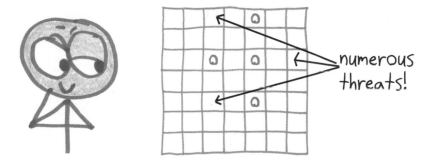

In short, Corners is like wine, presenting a balance of distinct flavors. Also like wine, it's a bit of an acquired taste. At first, threats and opportunities may lay hidden in plain sight. You scan the field of dots like it's one of those old Magic Eye patterns, a psychedelic wallpaper that refuses to resolve into the advertised image. The soundtrack of your early matches will be "Oh no, I didn't see that!" and "Wait, how did I miss the threat?"

But give it time. Soon, squares will pop out unbidden. To learn Corners is to learn a whole new way of seeing. In that sense, Corners is like every game: a training ground for our perception.

WHERE IT COMES FROM

Corners' most direct predecessor is Walter Joris's game Territoria. I loved that game's mechanic (whereby completing a square lets you fill its interior), but I found my matches ending as runaway victories or densely tangled draws, so I introduced the "empty dot vs. filled dot" distinction, plus the step of "claiming" a square.

Joris's game, in turn, echoes a variety of mathematical precedents, all sharing a common theme: They ask you to spot potential corners in an array of dots, to extract a simple pattern from a field of distractions.

First are visual riddles called Zukei (roughly translated as "looking for shapes"), created by the puzzle master Naoki Inaba. In each, your task is to find a specific shape among a collection of dots, a bit like finding a constellation among stars.

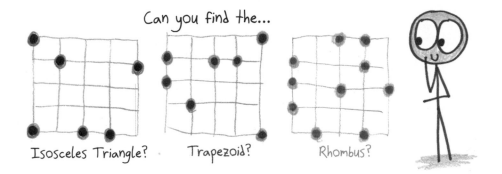

Can you find the...

Isosceles Triangle?   Trapezoid?   Rhombus?

A similar puzzle, from the site Play with Your Math by teachers Joey Kelly and CiCi Yu, asks the reverse question: How many X's can you place on a grid *without* creating any rectangles?

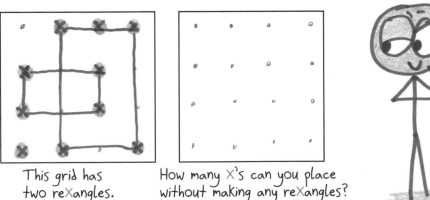

This grid has two reXangles.

How many X's can you place without making any reXangles?

Then there's this beauty, which inspired Kelly and Yu, and for which mathematicians spent two years searching. It's a 17-by-17 square, in four colors, with a very special property: No quartet of same-color dots ever forms the corners of a rectangle. To borrow Joey and CiCi's phrase, it's the anti-reXangle to end all reXangles.

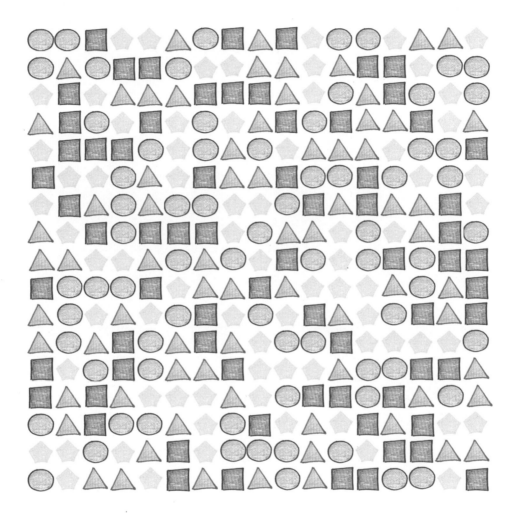

WHY IT MATTERS

Because games rewire our perception.

The first time you encounter a sudoku puzzle, it's a slog. You attend to one cell at a time, thinking, "Could this be a 1? What about a 2? Or a 3? Maybe a 4?" In one study, novices took 15 minutes to find an average of just two digits. That's enough time for an experienced player to solve a whole puzzle. The newbies' reasoning was sound, but their speed was agonizing.

By solving a few puzzles, you learn new and speedier ways to scan. At my own modest level of skill, I picture each 7 as "sweeping out" a whole row and column. This helps to find nooks where the next 7 must go.

Full-on experts seem to acquire a third, sudoku-specific eye. In one of my all-time favorite YouTube videos, sudoku master Simon Anthony confronts a fiendish puzzle with only two numbers given.

"He's got to be joking," Simon sighs.

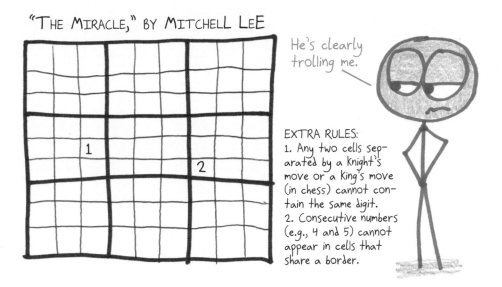

"THE MIRACLE," BY MITCHELL LEE

He's clearly trolling me.

EXTRA RULES:
1. Any two cells separated by a knight's move or a king's move (in chess) cannot contain the same digit.
2. Consecutive numbers (e.g., 4 and 5) cannot appear in cells that share a border.

But over the next 20 minutes, Simon puts on a masterclass in perception. No step in his logic is particularly alien or hard to follow. The magic is that, over and over again, his attention leads him to just the right place. Simon talks about cells "seeing" one another, which I love: Just as his eye perceives the numbers, they seem, in Simon's eyes, to perceive each other.

The same thing happens as you improve at Corners. A meaningless grid becomes a network of patterns and pressure points. An expert player is not a master thinker so much as a master perceiver, a masterful attender to the most promising possibilities and combinations.

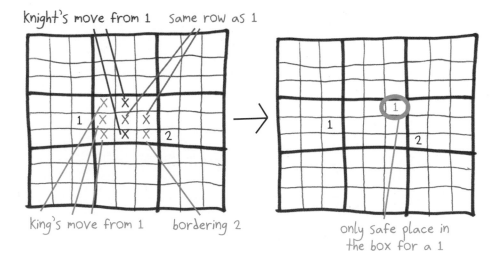

knight's move from 1   same row as 1

king's move from 1   bordering 2

only safe place in the box for a 1

A classic psychology study tackled a similar question: How do chess masters see the board? The masters were shown two kinds of scenarios: first, actual game situations, drawn from partway through real matches, and second, chaotic boards, in which the pieces had been distributed randomly across the squares, creating positions that violated the game's logic and rules.

How long would it take to memorize each kind of board?

Game Scenario   Randomized Board

As it turned out, the masters could memorize the game scenarios in mere seconds. But on the random configurations, they performed little better than novices.

Chess masters don't have "photographic" memory. If they did, they'd learn the chaotic boards as easily as the real ones. Instead, their memory's speed is the superficial consequence of a deeper power: structure. Their minds are multitiered storage systems for chess positions. Years of experience allow them to file away new game scenarios with ease. But this lends them no special insight into the chaotic scenarios. Their perceptual training is profound, yet very specific.

I suppose it's time to cycle back to the puzzle that opened this chapter: creating different sizes of squares on a 49-dot grid. The first possibilities you'll spot will be those aligned to the grid, like these:

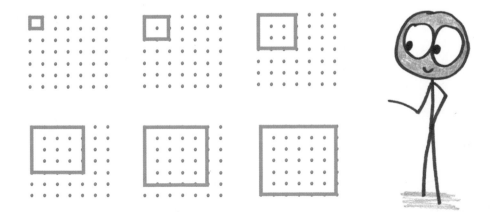

Keep looking, and you may notice those tilted at a 45° angle, so-called diamond shapes:

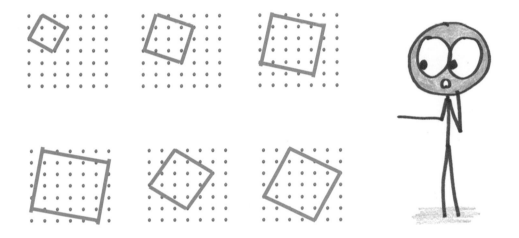

Then, rotating the book and giving it a healthy squint may reveal squares at stranger angles:

I excluded these oblique squares from Corners; the game felt complicated enough. But who knows? Perhaps my perception just awaits further training.

VARIATIONS AND RELATED GAMES

MULTIPLAYER CORNERS: For three or four players, use a larger board (such as 9 by 9). I also recommend "snaking" the order of turns: A, B, C, C, B, A, A, B, C, and so on.

PERIMETER-STYLE CORNERS: If you feel the game has too much of a "runaway winner" phenomenon, you can weaken the power of claiming a square. Instead of getting to place empty squares throughout the whole interior, you may place them only around the perimeter.

QUADS AND QUASARS: Invented in 1979 by college student G. Keith Still, and shared in the pages of *Scientific American* in 1996 by mathematician Ian Stewart, the game resembles a version of Corners where the first square wins.

You play on an 11-by-11 grid, with the four corners removed. Each player begins with 20 pieces called quads (red for one player, black for the other; you can use pennies and nickels) and seven blocking pieces called quasars (white for both players; you can use dimes).

Take turns placing quads on the board. The goal is to create the four corners of a square, which can be any orientation: aligned with the grid, on the 45° diagonal, or at any oblique angle.

The quasars are just for blocking. You can play as many of them as you like on a turn (before then playing your quad), but you only get seven for the game, so deploy them carefully.

If the game ends without a square formed, the player with the most unplayed quasars wins.

AMAZONS

A GAME OF VANISHING TERRITORY

Like a butterfly, a Pokémon, or the Leonard Cohen song "Hallelujah," the game of Amazons needed a long time to reach its final form. The concept wound its way from a mathematical text (in the 1940s) to a *Scientific American* column (in the 1970s) to a German board game publisher (in the 1980s) and then at last to an Argentinian puzzle magazine (in the 1990s).

It was worth the wait. Amazons' fans consider it a masterpiece. Avid player Matt Rodda calls it the perfect midpoint between games like Teeko (newbie-friendly, but lacking depth) and ones like chess or go (deep, but bogged down with specific tactics to memorize). Amazons is the best of both worlds, deep yet accessible, a combination game *par excellence*.

HOW TO PLAY

What do you need? Two players, a chessboard, three black pieces, and three white pieces. The pieces are called Amazons and move like queens (but feel free to use whatever pieces are handy).

You also need a collection of coins or counters for marking "destroyed" squares. If you draw the 8-by-8 grid on paper, you can mark "destroyed" squares with a pen or pencil.

Either way, here's how the six pieces begin:

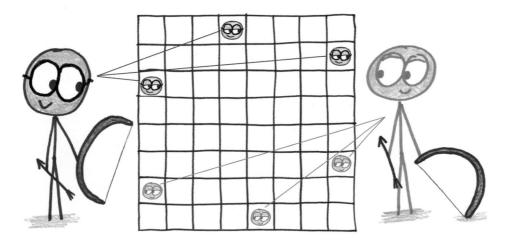

What's the goal? Have the last Amazon left standing when the board's destruction is complete.

What are the rules?

1. On each turn, select one of your Amazons, and move it as you would a queen in chess: **as far as you want in any direction**, including diagonals.

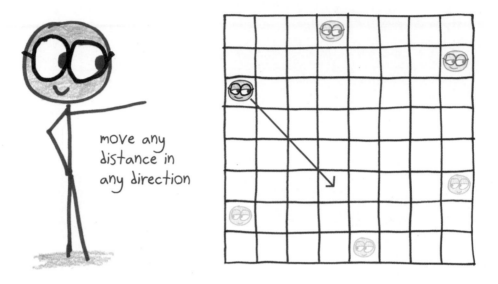

move any distance in any direction

After moving, **your Amazon fires off a "burning arrow" in any direction from its new location**. This arrow also moves like a queen. It **destroys the square where it lands**, leaving the squares it passed through unaffected.

fire an arrow any distance in any direction

2. Continue in this manner. **Destroyed squares and all pieces act as insurmountable obstacles.** Thus, an Amazon cannot move through them . . .

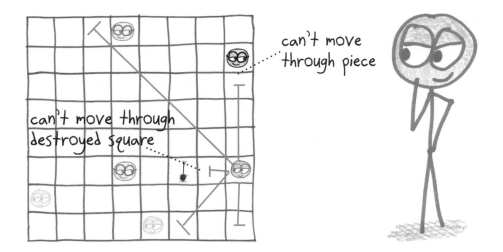

can't move
through piece

can't move through
destroyed square

. . . and **an arrow cannot be fired across them.**[7]

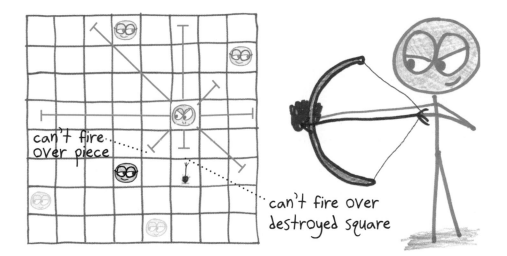

can't fire
over piece

can't fire over
destroyed square

3. Eventually, the entire board will be destroyed or unreachable, leaving
    all the Amazons trapped. Whoever has the **last Amazon able to move
    is the winner**.

---

7   Okay, yes, in real life arrows can go over things. But not here. If it helps, imagine that each arrow, upon
hitting the ground, creates a mile-high column of fire.

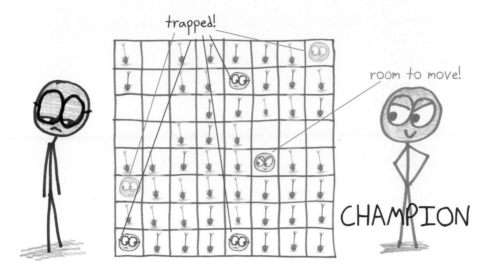

TASTING NOTES

On each and every turn, one square goes up in flame. The playing area thus transforms from a safe, wide-open continent into a fragile archipelago of shrinking islands until, in the end, all is consumed by apocalyptic fire.

If that doesn't sound fun, what does?

Toward the end, the board fragments into *royal chambers*: enclosed regions each containing only a single piece (or multiple pieces of the same color). From this point, the strategy is straightforward: Tour your chamber one square at a time, firing an arrow at each square as you leave it, thereby taking as many moves as possible to destroy your own refuge.

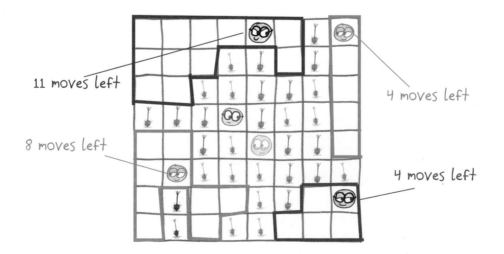

To win, you want to carve out spacious royal chambers for your own pieces, while trapping your opponent in tiny ones. Easier said than done. Gameplay exhibits frequent and sudden reversals of fate. One moment it seems you're about to trap your enemy in a suffocating chamber; the next, they've somehow slipped free, trapping you instead.

"Amazons is not a game of increments," reports one player. "Sudden unexpected moves can change the entire complexion of the game from that point forward."

WHERE IT COMES FROM

In the 1940s, David L. Silverman devised Amazons' grandparent: Quadraphage. The name comes from *quad* meaning "square" and *phage* meaning "eater."

In Quadraphage, a chess piece (for example, a king or a knight) attempts to escape off the edge of the board, while a square-eating enemy attempts to trap it by placing counters. It's a good mental workout, but it's less a game than a family of puzzles, each losing its fun when solved.

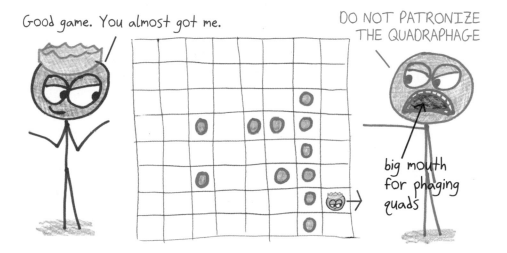

A proper game arrived in 1981, with Alex Randolph's Pferdeäppel. The name is German for "horse droppings" (a sentence I never expected to write). Two knights begin in opposite corners of the chessboard; on each turn, you move your knight in the way a chess knight moves (two squares horizontally and one vertically, or vice versa) and place a counter (symbolizing a fresh horse poop) on the square you just vacated. Knights can leap over poop, but must not land on it. Whoever captures their opponent (by landing on them) or leaves them trapped (with only poopy moves available) is the winner. The game is not to be played at mealtime.

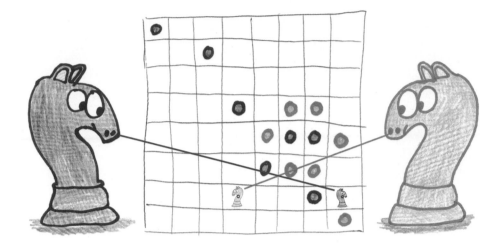

Finally, in 1992, Argentinian game designer Walter Zamkauskas published (in the puzzle magazine *El Acertijo*) a game he'd invented four years prior: Amazons. His masterstroke, in my view, was the rule for destroying squares. Where Quadraphage gives you a free choice and Pferdeäppel gives you no choice, Amazons gives you a cleverly constrained choice.

WHY IT MATTERS

Because Amazons showcases the essence of play: meaningful decisions.

Each move in Amazons combines two actions: slide a piece, then fire an arrow. Sounds humble enough, but don't be fooled. This simple combination yields staggering complexity, as captured by a concept called *branching factor*.

A game's branching factor answers the question: *On the average turn, how many options do you have?* Tic-tac-toe, for example, has a branching factor of about 5. Thus, on a typical turn, you're picking between five or so choices. Multiplying out all of the possibilities, you'll find that tic-tac-toe can unfold in more than 250,000 ways. Not bad for such a simple game.

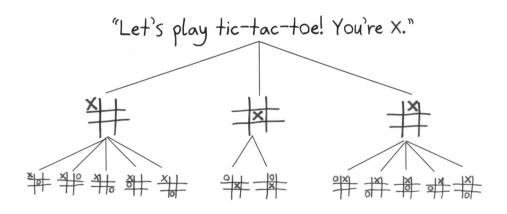

Chess's branching factor is estimated at 30 or 35, meaning that the typical move offers dozens of choices. That's pretty branchy. The first four moves (two per player) can unfold in over a million ways, and the entire game tree comprises roughly 10^{120} possible games. That's far greater than the number of subatomic particles in the visible universe.

Yet even chess cannot hold a candle to the branching factor of early-game Amazons. Your opening move gives more than 50 options of where to slide, each followed by at least 15 options of where to fire your arrow, for an eye-popping total of almost 1,000 possibilities. That's just the first move. Though the branching factor declines as the board fills up, it remains in the hundreds for many moves to come, outstripping chess or even go.

"The reason there are so many 'surprises' in Amazons," writes game designer Nick Bentley, "is that the branch factor is gigantic . . . [I]t's really easy to miss threats inside that bushy game tree."

In gaming circles, a bit of folk wisdom defines a game as "a series of interesting decisions." But a big branching factor is no guarantee of interesting decisions. Just as a game can bore you with a lack of options, it can paralyze you with an overabundance. Instead, I see two requirements for decisions to feel interesting: (1) The ability to tell **which decisions will bring progress** toward your goal; and, just as crucial (2) **gaps or flaws in this ability.**

Consider the classic game of Nim. You set up the table with a few piles, each containing some number of items (traditionally one, three, five, and seven). Then, on each turn, choose a pile, and remove as many items as you like—as few as a single item, or as many as the entire pile. Keep taking turns. Whoever removes the final item wins.

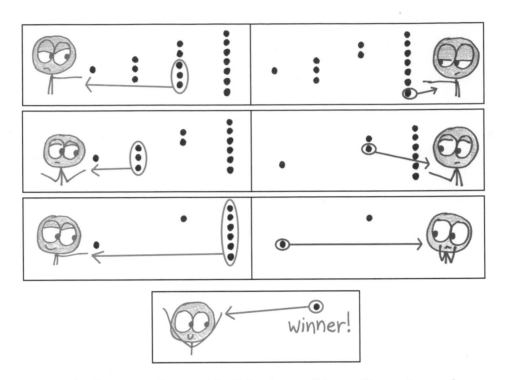

For the first several turns, it's all but impossible to tell a good move from a bad one. In this stage, Nim fails requirement #1. Later, with just a few items left, you can examine every possibility and determine the optimal choice. But that fails requirement #2. When the best choice is readily apparent, it's no choice at all. It's like Hein Donner said of chess: "Give me a difficult positional game, I will play it. But totally won positions, I cannot stand them."

In other words, Nim leaps straight from random moves to predetermined ones, from 100% ignorance to 100% certainty. It never passes through the middle ground of hunches and heuristics and partial knowledge, that misty realm where meaningful choices reside.

In short: Nim is a great piece of mathematics, but it's a lousy game.

How Confident Are You in Your Move?

0% the "too hard" kind of boring   the middle ground where good games live   the "too easy" kind of boring 100%

Amazons sits in that sacred middle space. Though you can never identify the best move, you'll quickly develop intuitions for good ones. Are my pieces mobile and unencumbered, able to reach any area of the board? Are my opponents' pieces hampered and clustered together, struggling to escape from fast-shrinking regions?

You won't always get it right. From time to time, you'll find yourself thwacked in the face by some unseen branch of that bushy game tree.

But hey, if that doesn't sound fun, what does?

VARIATIONS AND RELATED GAMES

6-BY-6 AMAZONS: For a quicker game, play on a 6-by-6 board with two Amazons each, in the starting position below.

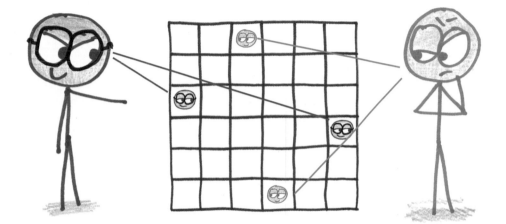

10-BY-10 AMAZONS: Longer and more involved than the 8-by-8 version, this is Amazons as Walter Zamkauskas originally published it. Begin with four Amazons each, positioned as below.

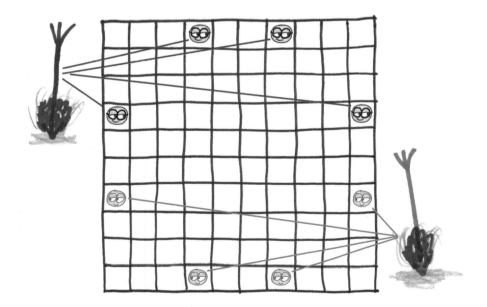

COLLECTOR: A nifty Amazons-like game designed by Walter Joris. Play on a 6-by-6 grid, and on each turn, (1) **mark any box you like**, and (2) **eliminate any empty neighboring box** (including diagonals). Play until no more moves are possible. The **winner is whoever creates the largest group of connected marks**. Diagonal connections count.

QUADRAPHAGE: The great-grandparent of Amazons. I recommend playing with a king, knight, rook, or queen.

1. The first player begins with a chess piece (let's say a king) anywhere on the chessboard. On each turn, they **move the piece as it would move in a game of chess**.
2. The second player begins with a handful of coins. On each turn, they **place a coin on a square of their choice**, thereby marking it off-limits to the first player.
3. The first player, **upon escaping off the edge of the board, scores 1 point per coin** that has been played. If trapped and unable to escape, they score 0 points.
4. After playing, **switch roles**. Whoever scores more points is the winner.

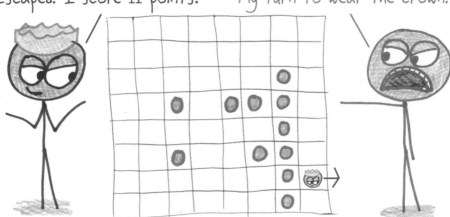

PFERDEÄPPEL: Each player controls a **knight**, which moves in its usual manner. The knights begin in **opposite corners** of an 8-by-8 chessboard. On each turn, **move your knight and place a marker on the square you just vacated**, signifying that it can never be occupied again. Whoever **captures the opposing knight** (by landing on it) **or traps it** (by leaving it unable to move) is the winner.

A COMBINATION PLATTER OF GAMES

The games ahead may feel a bit eclectic. They involve placing dominos, connecting lines, drawing X's, and rotating objects. "What's the deal?" you may ask. "Is *everything* a combination game?"

Well, yeah. Consider the wisdom of the eighth-century Hebrew text *Sefer Yetzirah*. It says that everything, all of reality, is made from combining and recombining the letters of the Hebrew alphabet. "God drew [the letters] . . . combined them, weighed them, interchanged them, and through them produced the whole creation and everything that is destined to be created." In this view, God is a combinatorist, and you and I are mere combinations.

COMBINATORIAL CENSUS

RACE/ETHNICITY: I am a vast combination of...

☐ cells
☐ molecules
☐ atoms
☐ Hebrew letters
☐ I prefer not to say

So, my fellow combination of molecules, I hope you enjoy these little combinations of rules.

TURNING POINTS

A GAME OF RUNAWAY ROTATIONS

For this dizzying two-player game, you'll need a square board (I suggest 4 by 4 for a quick game, or 6 by 6 for a longer one), plus a bunch of **movable pieces that can face in a specific direction**. I enjoy using Goldfish crackers, while Joe Kisenwether, the game's creator, recommends drawing arrows on cardboard poker chips.

Sit on opposite sides of the board, and **take turns placing pieces** on empty squares. Each piece must face one of its four neighboring squares. (Diagonals don't count.) If this **neighboring square is occupied**, then **the piece on it rotates 90° clockwise**. If the rotated piece is now pointing at another piece, then that one also rotates 90° clockwise, and so on, until you reach a piece that points to an empty space (or to the edge of the board).

empty square; turn ends

Play until the board is full, and then **score 1 point per piece pointing toward your side** of the board. The highest score wins. To play with four players, use all four sides of the board. To play with three or six, use a hexagonal board made of smaller hexagons, and 60° rotations instead of 90° ones.

DOMINEERING

A GAME OF CROWDING DOMINOS

In this game, two players take turns placing dominos on a rectangular grid. **One player lays dominos vertically; the other, horizontally.** (You can ignore the numbers on the dominos.) If it is your turn and you have **nowhere to place a domino, then you lose**.

The early moves feel a bit random. But soon, corridors start to appear. You and your opponent vie to secure "safe" spots for the future. Eventually, the board breaks down into disconnected chunks, and you can tally exactly how many moves each player has left.

contentious zone

WINNER

vertical:
2 safe moves

horizontal:
3 safe moves

Also known as Stop-Gate or Cross-Cram, the game is a classic of combinatorial game theory, prominently featured in the canonical text *Winning Ways for Your Mathematical Plays*. Whereas other classics (such as the million variations on Nim) serve better for mathematical analysis than casual gameplay, I find that Domineering works on both levels.

By the way, no need for actual dominos. You can **play by filling in squares on a paper grid**.

HOLD THAT LINE

A GAME OF SNAKING GROWTH

Sid Sackson devised this game as an alternative to tic-tac-toe. "If all the people who ever played Tic-Tac-Toe were laid end to end," he wrote, "they would promptly fall asleep." He hoped to replace those drawish doldrums with a more flavorful game, one that never ends in a tie.

To begin, draw a 4-by-4 array of dots. The **first player connects any two dots with a straight line of any length**, which may be vertical, horizontal, or 45° diagonal.

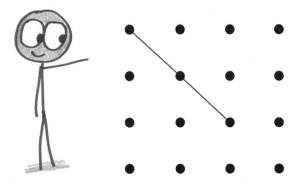

Then, **take turns extending this line from either end** by drawing another line (vertical, horizontal, or 45° diagonal). Extensions may be any length, but may not cross or touch. Play until no further extensions are possible. The **person to draw the last extension is the loser**.

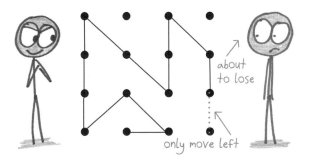

Sid's game resembles an older one published by Édouard Lucas alongside Dots and Boxes. Édouard's version has just a few differences: (1) You play on a 6-by-6 array of dots. (2) Each move must be a short vertical or horizontal line, connecting two adjacent dots. (3) You must build off of your opponent's most recent move, meaning that the "snake" grows from only one end. (4) The last player to move is the winner.

CATS AND DOGS

A GAME OF REFUSING TO GET ALONG

Draw a 7-by-7 grid on paper. Then, **take turns placing your respective animals**: for one player, "cats" (i.e., X's), and for the other, "dogs" (i.e., O's). **Cats and dogs must never occupy neighboring squares**, not even diagonally. **The last player able to move is the winner.**[8]

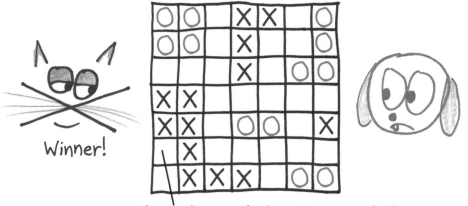

Winner!

two cat moves left; no dog moves left

The game was developed by algebraist Simon Norton and was known in his honor as "Snort." Instead of cats and dogs, he imagined bulls and cows, grazing in various fields, liable to noisy snorts of distraction if placed too close to the opposite sex.[9] The fields need not follow a grid arrangement; you can draw any convoluted map of regions that you like.

A related classic of combinatorial game theory, Col, reverses the key rule: *Opposite*-species neighbors are allowed, and *same*-species neighbors are forbidden. Col is easier to analyze mathematically and, perhaps for that reason, less fun to play. Scattering your cats around the board, trying to keep them apart? Meh. Fencing off safe territories with jagged walls of cats? Now *that's* fun.

---

8   Just one extra technicality: The first player's first move cannot be in the very center square.
9   The "cats and dogs" theme comes from a lovely wooden version by the Portuguese company LuduScience.

ROW CALL

A GAME OF SHARED CONTROL

It's tic-tac-toe with a simple twist: You don't fully control where your mark goes. Instead, on each turn, **you pick a row or column**, and your opponent makes the final decision of **where in that row or column to place your symbol**. Elise Johnson-Dreyer's students dubbed it Bossy Tic-Tac-Toe, which I like, because it's unclear to me who the boss is.

Play on a 4-by-4 grid, with the columns (Y-O-U-R) and rows (P-I-C-K) labeled for ease of reference. **Whoever gets three in a row first is the winner.**

playing O   playing X

Early in the game, it feels like your opponent wields more control over your moves than you do. But as the game goes on, the negotiating power shifts. Sometimes, you can pick a row with only one spot remaining, leaving your opponent no choice of where to place your mark. For a longer game (with an admittedly slower start), play on a 5-by-5 grid, aiming to get four in a row.

It's also fun to apply this "you finalize my move" principle to other games, such as Dots and Boxes (I pick a row of dots, you pick where to place my mark) or chess (I pick one of my pieces, you decide where to move, with the caveat that if I say "check" or "capture," then you must do so).

IV

GAMES OF RISK AND REWARD

YOU PLAY GAMES of risk and reward every day. Crossing against a DON'T WALK sign; correcting your boss in public; drinking expired milk—all of these "games" promise rewards, such as milk, yet carry risks, such as milk poisoning. Mathematics has a lot to say about such decisions.

Here, turn on the TV, and I'll show you.

Lots of television shows are nothing but extended case studies in the mathematics of risk and reward. Some pose classic puzzles from **probability theory**, a branch of mathematics that quantifies uncertainty. Others deal in **game theory**, the mathematics of strategic interactions. Many shows involve both, interwoven to create vexing mathematical dilemmas.

No, I'm not talking about BBC specials starring Hannah Fry. I'm talking about game shows.

Take *Deal or No Deal*. It begins with a man named Howie Mandel striding across a dark stage, saying "one million dollars" over and over, as if summoning an ancient spirit. Also onstage are 26 closed cases, held by 26 women in identical dresses.[1] Each case contains a different amount of money, from $0.01 to—say it with me, Howie—$1 million.

---

1 I've always assumed they are bridesmaids on their way to an unusually large wedding.

The contestant selects a case. The other cases are then opened, one by one, each eliminating a prize from the list of possibilities. Every so often, a shadowy banker phones in an offer to buy the contestant's case.

Help us, probability theory! What should the contestant in this scenario do?

BANKER'S OFFER: $77,000	PRIZES REMAINING	
	0.01	1,000
	1	5,000
	5	10,000
	10	25,000
	25	50,000
	50	75,000
	75	100,000
	100	200,00
	200	300,000
	300	400,000
	400	500,000
	500	750,000
	750	1,000,000

DEAL... or NO DEAL?!

First, let's imagine playing the game to its conclusion a million times over, the contestant's case revealing a different value each time. Nine values remain, so in ⅑ of the games, it's a sad $75. In another ⅑, it's a cool half million. And in the other ⅞, it's all those values in between. Calculate the average of all such values: That's the case's "expected" value. It captures, as the scholar Christiaan Huygens put it in the 1600s, "for what price I would reasonably cede my game to another who would desire to continue in my place."

Problem solved, with just one tiny caveat: On *Deal or No Deal*, the banker never seems to offer the expected value.

$77,000? That's way below expected value.

You mean... one million dollars?

Howie, be serious. It's $107,000 or so. How could expected value exceed the maximum value?

PRIZES REMAINING	
0.01	1,000
1	5,000
5	10,000
10	25,000
25	50,000
50	75,000
75	100,000
100	200,00
200	300,000
300	400,000
400	500,000
500	750,000
750	1,000,000

It's not that the banker (really an algorithm controlled by the show's producers) wants to save money. The true goal is to maximize viewership by creating dramatic and difficult choices. Lowball early offers (often less than 40% of expected value) encourage contestants to keep playing. Later offers get better, but never reach expected value, because that'd be too easy: Most folks are risk-averse, and would cheerfully take a guaranteed $500,000 over a coin flip for $1,000,000. The question is whether they would take $400,000, or $300,000. That's a harder choice, and thus better television.

But enough of Howie and his million dollars. Let's switch over to a calmer, more dignified, more enlightened game show: the trivia classic *Jeopardy!*

Every episode ends with a high-stakes question called Final Jeopardy. Contestants get to place secret wagers on their own answers, gambling their winnings to date on the as-yet-unseen question. Most nights, the wagers follow a reliable pattern. The contestant in second (the challenger) wagers everything, while the contestant in first (the leader) wagers just enough to exceed that total by $1.

Strange, isn't it? Though they're given secrecy, the contestants almost never exploit the element of surprise, preferring instead to make the predictable choice. John von Neumann would not be impressed. "Real life consists of bluffing," he once said, "of little tactics of deception, of asking yourself what is the other man going to think I mean to do. And that is what games are about in my theory."

He's referring to game theory, the mathematics of strategic interactions. You'll explore von Neumann's legacy firsthand in this section, as games like Undercut and Paper Boxing force you to psychoanalyze your opponent, anticipating their moves to stay one step ahead. Your optimal choices will depend intimately on theirs.

So how does this play out in Final Jeopardy?

Well, under the standard wagers, the challenger is left hoping for a specific scenario: to get the question right while the leader gets it wrong.[2] Any other outcome, and the leader triumphs.

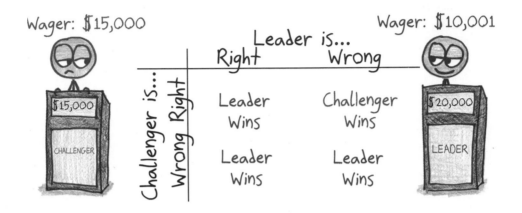

Is this the best the challenger can do? Not at all. By betting $0 instead, the challenger would win whenever the leader slips up—even if the challenger answers wrong, too.

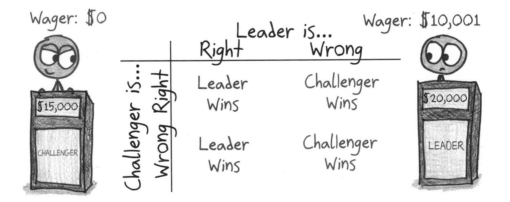

But that move carries a risk. What if the leader anticipates the challenger's sneaky strategy? Then the leader can bet $0, too, and thereby guarantee victory.

---

2   Coming in first is more important than maximizing your winnings, because the first-place finisher keeps their whole winnings and plays again on the next show, while the second-place finisher goes home with $2,000.

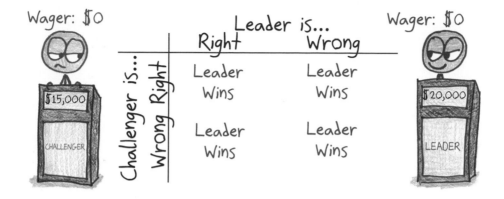

Then again, the challenger might anticipate this, and seize control by making a real bet.

Interesting, isn't it? We could run a similar analysis of almost any show you pick. When to buy a vowel in *Wheel of Fortune*; whether to venture a guess on *Who Wants to Be a Millionaire?*; what price is right on *The Price Is Right*. Each is a matter of game theory and probability theory, a question of calculating risks and rewards.

These silly games are the model systems by which we understand reality.

Here's a spooky case in point. Although John von Neumann first developed game theory to analyze his neighborhood poker game, he soon realized it was an uncanny fit for geopolitics. In the strategic interactions of the Cold War, von Neumann glimpsed a simple payoff matrix. The US and USSR both desired control of the globe. If both sought peace, there'd be peace. If one sought peace while the other attacked, the aggressor would triumph. And if both attacked, there'd be nuclear war.

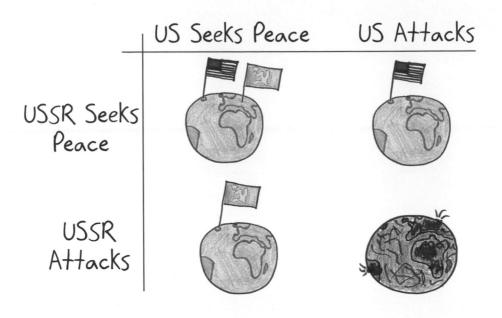

He concluded that the US had only one logical choice: a public and unwavering commitment to retaliatory violence. Attack us, and we shall end you, even if that means ending ourselves. Thus was born the uneasy peace of mutually assured destruction, aptly known as MAD.

The world riding on the logic of a game.

Lucky for us, the risks and rewards in this section don't carry quite the same stakes. Try to remember that when your 11-year-old acquaintances are humiliating you at Undercut. The world isn't ending in a nuclear fireball; you just kind of wish it were.

UNDERCUT

A MIND GAME—OR, FOR REAL EXPERTS, A MINDLESS ONE

Undercut is the perfect game for a family road trip. All you need are your hands, your wits, and an opponent you enjoy infuriating. When I first taught Undercut to 11- and 12-year-olds, they loved it, not least because they thrashed me every time. I came home licking my wounds and challenged my wife to a match of redemption. Then, after I undercut her three times in a row, she fixed me with a glare and extracted a promise that we'd never, ever play again. Anyway, with this game now banned in my household, I offer it as a gift to yours.

HOW TO PLAY

What do you need? Two players, each with five fingers.[3] You may also want pencil and paper to keep score.

What's the goal? Pick a number that is 2, 3, or 4 greater than your opponent's—or better yet, exactly 1 smaller.

What are the rules?

1. Each player **thinks of a number from 1 to 5** and then, on the count of three, reveals it. **You score 1 point per finger.** Simple enough, but there's one crucial exception . . .

---

3   If one of you has more fingers than the other, please agree upon a number of usable fingers in advance.

2. **If your number is exactly 1 larger than your opponent's, then you've been undercut,** and your would-be points go to your opponent instead.

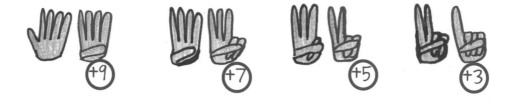

3. The game continues, round by round, **until one player pulls ahead by 11 or more points,** at which time they are declared the winner.

Undercutter
Extraordinaire

Overcutter
Ordinaire

TASTING NOTES

Sweet and salty in rapid succession, Undercut is less a math game than a mind game.

Say you're expecting me to play a 4. You'll therefore want to play a 3 . . . which means I should actually play a 2 . . . unless you've anticipated this maneuver, and thus choose 5 . . . in which case I should revert to the 4 you were originally expecting . . . and so on and so on, deep into the thicket. I liken Undercut to that classic scene in *The Princess Bride* where the Sicilian faces two cups of wine—one poisoned, the other not—and talks himself in desperate circles, trying to deduce which cup is which. Undercut is like that, except with less poison. Which is honestly not a bad blurb. *Undercut: A swig from a poison chalice, but without the poison.*[4]

WHERE IT COMES FROM

In the summer of 1962, math students Douglas Hofstadter and Robert Boeninger were winding their way through the forests of southern Germany, on a bus bound for Prague, bored out of their skulls. To kill the time, they devised Undercut, and proceeded to play round after round, like rams locking horns, as the miles sped by outside.[5]

Later that autumn, Hofstadter whipped up an Undercut-playing computer program. It used numbers instead of fingers—that is, digits instead of digits. Its aim: to detect and exploit patterns in the opponent's moves.

"My program often started out on a losing track," Hofstadter later wrote in *Scientific American*, because "it had not yet 'smelled' any patterns in the opposing program's behavior." Eventually, though, it would "catch the scent" of the opponent's thinking and surge to victory, undercutting like a swift-fingered samurai. Hofstadter recalls "a feeling of overwhelming power." But just as his mad-scientist cackles began to crescendo, a challenger emerged.

His name: Jon Peterson. His program: a simple application of game theory.

"It's not that my program got trounced by his," Hofstadter explained. "It just never caught on to any patterns." No matter how long the two programs played, they just seesawed back and forth, settling into an eternal draw. The bloodhound could catch no scent. "Baffling," Hofstadter wrote.

Baffling, that is, until Peterson explained how his program worked. It simply ignored Hofstadter's, and made random choices, governed by a special set of probabilities.

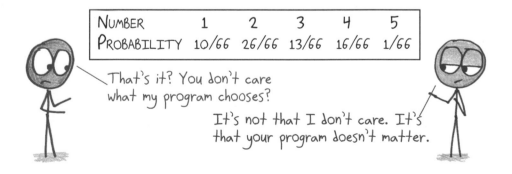

NUMBER	1	2	3	4	5
PROBABILITY	10/66	26/66	13/66	16/66	1/66

That's it? You don't care what my program chooses?

It's not that I don't care. It's that your program doesn't matter.

---

4 I read this to my wife, who replied, "It's more poisonous than you're making it sound." So, still banned, I guess.
5 In the original version, one player picked numbers from 1 to 5, and the other from 2 to 6. Although they didn't realize it, the latter player has an advantage of 0.28 points per round (assuming both players apply the game theoretic solution). Just as well they switched to the symmetric version.

In effect, Peterson's program was rolling a 66-sided die, with 10 sides labeled "1," 26 labeled "2," 13 labeled "3," 16 labeled "4," and the final side labeled "5." This rigid, thoughtless strategy achieved a naïve and maddening brilliance.

It could not beat you. But by the same token, it could not be beat.

Say you decide to play a 4. In fact, you might as well play it 66 times in a row, because Peterson's program, as dynamic and responsive as a brick, will simply roll its virtual die over and over, oblivious to your strategy (or lack thereof).

Thus, at the end of round 66, you'll wind up with results something like this:

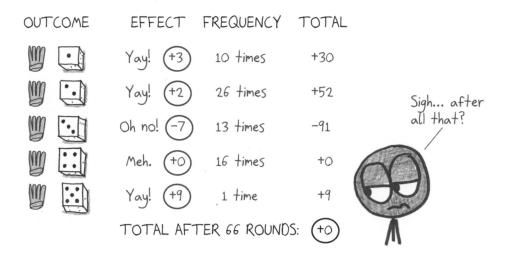

OUTCOME	EFFECT	FREQUENCY	TOTAL
✋ ⚀	Yay! (+3)	10 times	+30
✋ ⚁	Yay! (+2)	26 times	+52
✋ ⚂	Oh no! (−7)	13 times	−91
✋ ⚃	Meh. (+0)	16 times	+0
✋ ⚄	Yay! (+9)	1 time	+9

TOTAL AFTER 66 ROUNDS: (+0)

Sigh... after all that?

Five and a half dozen rounds later, where do you stand? On average, tied. You can't beat Peterson's program by playing 4. Nor can you lose. In the long run, you can only break even. The same is true of 1, 2, 3, and 5: no matter what number you throw at this 66-headed monster, it will fight you to a standstill, never pulling far ahead, never falling far behind, everything canceling out in a wash, as if you were racing your own shadow.

Hofstadter seethed. He called it "a humiliating and infuriating experience." After all, there's nobility in failing against an extra-human intelligence such as chess's Deep Blue, go's AlphaGo, or soccer's Megan Rapinoe. But against a random number generator? That's just sad.

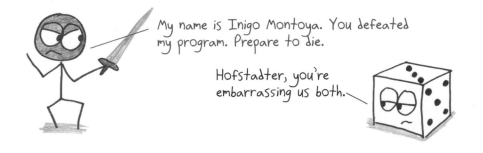

My name is Inigo Montoya. You defeated my program. Prepare to die.

Hofstadter, you're embarrassing us both.

Sad, but inevitable. You can't outguess randomness, and randomness can't outguess you.

WHY IT MATTERS

Because randomness is an overpowering strategic tool—a tool that we're constitutionally incapable of wielding.

Say you're a top-tier baseball pitcher. You have several weapons in your arsenal: fastball, curveball, knuckleball, matzoball. How do you pick which one to throw? If you want to baffle the batter, it's best to give no clues, to obey no rules—in short, to randomize.

Or say you're a Naskapi forager, hunting caribou. Where should you seek out the herd? Fall into a pattern—visiting locations in a regular cycle, or always revisiting the last successful site—and the caribou may learn to avoid you. The best bet is to randomize.

Or say you're an ancient Roman general, planning when and where to attack. You don't want your enemy to anticipate your moves, do you? Only one choice, then: randomize.

A final example, for your irredeemably modern self: picking a password. You don't want something trite and guessable, like "MyDogIsCute" or "GO_YANKEES" or "passw0rd1234." Instead, you want to reach your hand into the giant hat of all possible passwords, and pick one out at . . . well, at random.

THE MANY USES OF RANDOMNESS

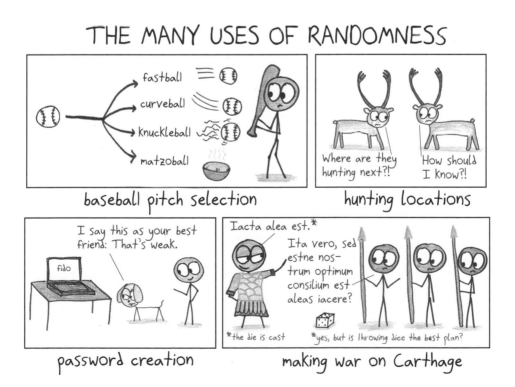

baseball pitch selection

hunting locations

password creation

making war on Carthage

A random strategy is both (a) utterly transparent and (b) utterly inscrutable. Your opponent knows exactly what you're up to, yet cannot outwit or outstrategize you, because you have forsaken the whole idea of wit and strategy, in favor of Zenlike surrender to chaos. It's brilliant. It's unassailable. There's just one problem.

We're constitutionally incapable of it.

Our brains aren't decks of cards. No, they're more like conspiracy walls, covered in newspaper clippings and red thread. We see patterns everywhere, whether or not they exist: cuddly animals in the shapes of clouds, meaningful trends in the graphs of stock prices, religious figures in toasted bread. Some folks even see logic in the final season of *Game of Thrones*.

No surprise, then, that we're lousy at acting randomly.

Name a random number from 0 to 9. Hey, did you pick 7? If so, you're not alone. It's the most common reply, drawing more than 30% of responses in one classic study. Which is strange, because there shouldn't *be* a most common reply, should there?

Responses to "Choose a Random Number"

actual chance level

LOOK! i'M So rANdoM!

You keep saying that. It feels highly nonrandom.

In a similar vein, if you flip 100 coins, then make up 100 fake flips off the top of your head, any decent computer will be able to tell them apart. The real results will have some long streaks, perhaps six heads or seven tails in a row. The fake ones won't.

random flips   faked flips

Physics professor Scott Aaronson once asked students to repeatedly type *f* or *d*, in order to see if a crude pattern finder could predict the next keystroke. The algorithm was simple: Just look at the most recent five-letter sequence (e.g., "ffddf"), scan for past instances, and guess whichever letter typically came next.

The pattern sniffer was right more than 70% of the time. "I couldn't even beat my own program," Aaronson wrote, "knowing exactly how it worked."[6]

Games lay bare our incapacity for randomness. In rock-paper-scissors, for example, the unbeatable strategy is to play each option at random, one third of the time. Scissors. Rock. Rock. Paper. Rock. Paper. Rock. Paper. Paper. Scissors.

Easy enough in principle. We just can't do it.

Masters of the game—yes, they exist—have observed persistent patterns. First, rookies overplay rock. Second, few people will choose the same symbol

6   Only one student seemed able to defy expectations, matching the prediction just 50% of the time. "We asked him what his secret was," writes Aaronson, "and he responded that he 'just used his free will.'" Good pro tip.

three times in a row. Third, someone who loses (say, scissors to rock) will often switch to the option that would've won (in this case, paper).

To achieve randomness—or anything close to it—we need to escape our own minds.

In building his Undercut champion, Jon Peterson didn't try to think up random choices all by himself. If he did, then he'd have settled into some pattern or another, and Hofstadter's program would have pounced. Instead, Peterson outsourced the randomization to the computer.[7]

Others have relied on nature. Roman generals often chose their moment of attack via bird augury, sacred signs drawn from the perching and calling of birds. Naskapi hunters, meanwhile, heat a caribou's shoulder blade over hot coals, then read the patterns of cracks and burns like a map to dictate the location of the next hunt.

Call it divine guidance. Call it pseudorandom number generation. The point is that, to escape our own ruts and cognitive biases, humans need help.

If you play enough Undercut, you'll come to see randomness as a kind of conservatism. Approach the game like a 66-sided die, and you'll neither lose big nor win big. Randomness creates a safe and stable perch, an endless tie, from which no one can dislodge you.

But if you want to do better than a perpetual draw, you've got to descend into the scrum. You've got to chase patterns in your opponent's play, thus allowing patterns to creep into your own. To wage war means opening up vulnerabilities in your own defenses.

To grasp victory, you must risk defeat.

Is this a wise choice? Well, unless you're a master of bird augury, it's not a choice at all. Sure, randomness may be the ideal strategy. But humans, bless us, have no knack for the ideal.

---

7   Funny story in this vein: Computer programmer Nick Merrill built an online version of Scott Aaronson's "f or d" predictor. Originally, it would keep you apprised of its predictive accuracy thus far, e.g., "71.59% accurate." But this created a vulnerability: By assigning even numbers to "d," and odd ones to "f," you could use the last digit of the accuracy report to generate random choices! Aaronson dubbed this a "security flaw."

VARIATIONS AND RELATED GAMES

FLAUNT: A juiced-up variant from Douglas Hofstadter. The rules are the same as Undercut, except that you score extra by playing the same number multiple times in a row. For example, if you play two consecutive 4s, the second one scores $4 \times 4 = 16$ points. Play another 4 immediately thereafter, and it'll score $4 \times 4 \times 4 = 64$ points. And so on.

But if you play a fourth consecutive 4 and are undercut by your opponent's 3, then your opponent will score their 3 points plus your $4 \times 4 \times 4 \times 4$, for a whopping total of 259.

Play until one person pulls ahead by a predetermined amount (e.g., 100 or 500).

MORRA: The lifestyle show *Euromaxx* calls it "the world's loudest game" (evidently having never heard my daughter's game "screechy fun times"). Anyway, it's a millennia-old Mediterranean pastime, devised by ancient Egyptians, enjoyed by ancient Romans, and described by one player as "a feeling, a passion, and a nation's culture."

On the count of three, you hold up a number of fingers from 1 to 5. At the same moment, you shout a prediction of how many fingers you and your opponent will hold up in total. A correct guess wins the round. If both players are wrong (or both right), you count to three and throw again, not pausing between rounds, until someone names the correct sum.

MULTIPLAYER UNDERCUT: When I shared Undercut with middle schoolers, they devised a multiplayer variant that I like even better than the original. The first to 30 points (or some other agreed-upon target) is the winner. Having several players opens up some exciting scenarios.

First, Abby may undercut Nathan, while LaRon is unaffected.

Second, LaRon may undercut Abby and Nathan simultaneously, stealing both their points.

Third, Abby and LaRon may both undercut Nathan, which results in their splitting his points.

Fourth—and most exciting of all—LaRon may undercut Abby, but be undercut himself by Nathan, who thus winds up with all the points.

UNDERWHELM: Another variant from Hofstadter, which he describes as "a tipped-over version of Undercut." Both players think of an integer, anywhere from 1 to infinity. Whoever names the lower number (e.g., 17 vs. 92) scores that many points (in this case, 17). Whoever names the higher number scores nothing.

There's one exception: If the numbers differ by exactly 1 (e.g., 24 and 25), then the player who named the *higher* number scores the sum of the two (in this case, 49).

Play to a predetermined total, such as 500, using pen and paper to keep score.[8]

---

8   Game theory yields a surprising optimal strategy, which I'll go ahead and spoil:

Number	1	2	3	4	5	6 and up
Probability	$\frac{25}{101}$	$\frac{19}{101}$	$\frac{27}{101}$	$\frac{16}{101}$	$\frac{14}{101}$	Never

Weird, right? Given an infinite menu, you always order one of the same five dishes. Still, in practice, neither player will have the patience for a tedious 2-and-3-at-a-time crawl to the finish line, so you're likely to deviate in psychologically interesting ways.

ARPEGGIOS

A GAME OF ASCENT AND DESCENT

Arpeggios is a lot like life: Luck is a big factor, but it's not everything. Choices matter.

For example, you will at some point roll dice that are mediocre for your purposes, but excellent for your opponent's. You now face a choice. Do you use the dice yourself, hampering your own progress to spite your rival? Or do you pass on the dice, thereby delivering gold to your sworn enemy? Depending on how you frame the question, you'll feel pulled in different directions.

Therein lies another resemblance between Arpeggios and life. When weighing risks and rewards, the answer is always shaped by how we ask the question.

HOW TO PLAY

What do you need? Two players: one "ascending," one "descending." (To assign roles, roll the dice once each: Lower number is ascender, higher is descender.) Also: pencil, paper, and a pair of standard six-sided dice. (These are easy to simulate; just do an internet search for "roll dice.")

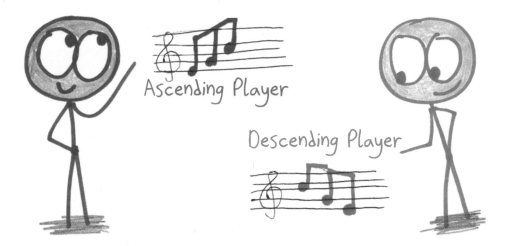

Ascending Player

Descending Player

What's the goal? Create a list of 10 numbers in ascending order (if you're the ascender) or descending order (if you're the descender).

You're a liar. What's the *actual* goal? Okay, fine. If you're the ascender, then your list doesn't have to ascend at *every* step. Once per game, you may break your pattern, and descend. But only once! Then you must return to ascending. (For the descender, this applies in reverse.)

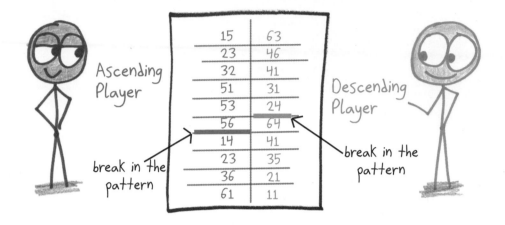

What are the rules?

1. To begin, the ascender rolls the dice. **Each die represents a digit**, and they can be combined in either order, to make a two-digit number.

2. Now, the ascender may either (a) **pick one of those two-digit numbers and place it in the next spot on their list**, or (b) **say "Pass."**

3. If they pass, then the **descender may steal the dice**; if so, this counts as their turn, and the next roll is again the ascender's. The descender may also **reject the passed dice**, in which case, they get to roll.

4. All turns unfold in this same manner. First, roll the dice. Then, either **use them or pass**. If you pass, your opponent may **steal them or decline** in favor of taking their own turn.

5. **If you roll doubles**, there's an extra option: **You may, if you like, reroll one die**, while leaving the other as is, before deciding whether to pass. Only one such reroll is allowed per turn (i.e., you may not re-reroll, even if your reroll gives the same number).

6. **Once per game—only once!—you may break your ascending or descending pattern**, as a kind of reset button. No need to announce this in advance; you can wait until you see the available dice.

7. The winner is the **first player to list 10 numbers**. Note that a repeated number (e.g., 41 followed immediately by 41) is not allowed.

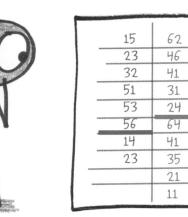

CHAMPION

15	62
23	46
32	41
51	31
53	24
56	64
14	41
23	35
	21
	11

TASTING NOTES

Arpeggios is a racing game. But it's a peculiar kind of race, in which rivals move in opposite directions, and your destination is the one place you must never reach. It's also a "roll and write" game. But it has little in common with Yahtzee (the genre's most famous example), nor does it quite resemble the recent titles that have swept the tabletop gaming world (Qwinto, Qwixx, Ganz Schön Clever, and so on). It's also a case study in risk and reward— even though, on most turns, the "right" choice will feel pretty clear. For all these reasons, I consider it the lemon-lime fizzy water of dice games: bubbly, refreshing, and a little hard to pin down.

WHERE IT COMES FROM

Walter Joris (my favorite mad visionary game inventor) created a simple concept called Pile Die. I added bells, whistles, player interaction, a second die, the ascending/descending distinction, and the titular musical theory reference (an arpeggio is a chord broken into a sequence of ascending or descending notes). Call it a basket for Orlin, with an assist from Joris.

WHY IT MATTERS

Because risks and rewards appear vastly different depending on how we frame them.

For example, let's say you're a doctor. Congrats on the MD, but don't get too giddy, because you face a grim task.

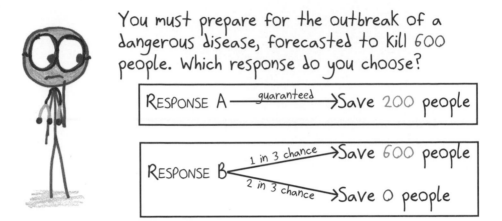

You must prepare for the outbreak of a dangerous disease, forecasted to kill 600 people. Which response do you choose?

RESPONSE A ——guaranteed——> Save 200 people

RESPONSE B —1 in 3 chance—> Save 600 people
          —2 in 3 chance—> Save 0 people

If you're like most folks—although, having received your medical licensure from a book of math games, you may not be—you go for response A. It feels reckless to gamble those 200 lives on the off chance of saving 400 more.

Alas, with that decision behind us, I have more bad news.

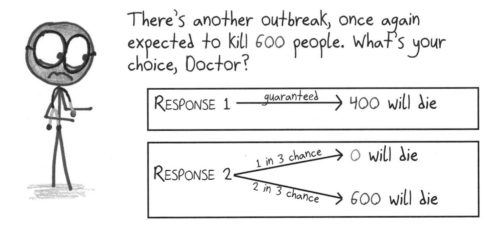

There's another outbreak, once again expected to kill 600 people. What's your choice, Doctor?

RESPONSE 1 ——guaranteed——> 400 will die

RESPONSE 2 —1 in 3 chance—> 0 will die
          —2 in 3 chance—> 600 will die

What's your choice now?

If you're like most people—though again, given your eventful medical career, you may not be—you go for response 2. It feels like an abdication of duty to let those 400 people perish. You've got to at least *try* to save them, even if it means putting another 200 people at risk.

Here's the problem: *The choices in each scenario are the same.* Response A is the same as response 1. Response B is the same as response 2. Only the phrasing differs. And yet, like a magnet held near a compass, the phrasing is enough to send our instincts spinning in different directions. Emphasize the 200 guaranteed survivors, and we shy away from risk. Emphasize the 400 guaranteed fatalities, and we leap at it.

Why should such a critical decision hinge on something as trivial as word choice?

The psychologist Daniel Kahneman, who devised this scenario with Amos Tversky, minces no words. He says it's because you're a moral nincompoop. "You have no compelling moral intuitions to guide you in solving that problem," Kahneman writes in *Thinking, Fast and Slow*. "Your moral feelings are attached to frames, to descriptions of reality rather than to reality itself." It's the ethical equivalent of picking politicians by their pantsuits.

On behalf of moral nincompoops everywhere: Ouch.

Yet, if I may contradict a Nobel laureate, I have a slightly cheerier take. Yes, framing is powerful. But it's a power we can harness. A good framing is a kind of illuminating paraphrase, able to transform confusion into clarity, a hopeless muddle into a helpful model.

Take, for example, my friend Adam Bildersee's deft reframing of Arpeggios. To begin, write a list of every possible number you might roll. Then repeat the list. We'll circle a number to show that you have selected it. For example, if you open the game with 16, 24, and 34:

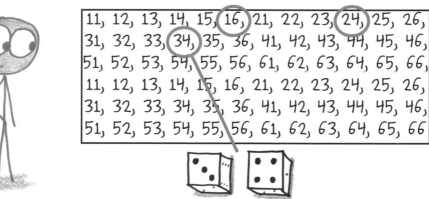

Picture this list as a kind of runway down which you advance, aiming to gather 10 numbers before reaching the runway's end. If you can claim a new number without consuming much runway, then it's a smart move.

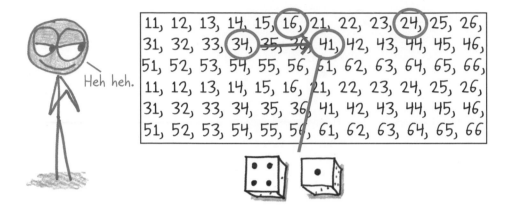

By contrast, using up a great deal of runway in a single turn is a dangerous risk. You might run out of space.

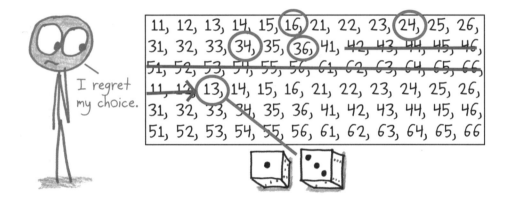

Here comes the surprising insight: the midgame "reset," when you descend for a single step rather than ascending, is not a special move at all. It simply occurs when you go from the third line of the runway (which ends with 66) to the fourth (which begins with 11). For example, going from 54 to 13 (thereby resetting) consumes the same amount of runway as going from 24 to 43 (an ordinary move).

On first encounter, the reset feels like a big red button, a spectacular once-per-game event. But it isn't. Framed appropriately, it's like any other move.

On First Encounter   After Reframing

Good framings often have that effect. What appears to be a cliff is revealed as a gradual hill. For example, consider one of the biggest questions a baby enthusiast like myself faces: *When should I get myself one of them babies?*

Delaying a few extra years has its rewards: more maturity, more financial stability, more hand-me-downs from friends with kids. But delaying also has its risks. In particular, the older you get, the harder it is to get pregnant.

Many medical experts have chosen a particular framing: They treat age 35 as a sharp cutoff. Before that, your womb radiates a golden light. Beyond that, you've reached "advanced maternal age" (previously known as "geriatric pregnancy") and all bets are off.[9]

But that's not the only framing. When economist Emily Oster dug into the research, she found no such cliff. Here's data from several thousand women in France who spent a year trying to get pregnant:

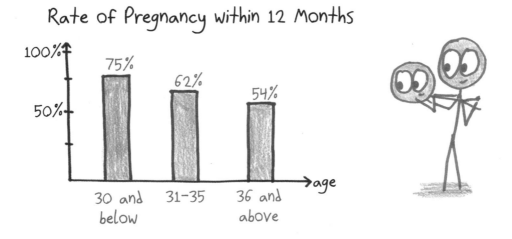

The bad news is that no age is 100% guaranteed. Starting when you're a teenager, every year brings a slight decline in fertility. The good news is that, even by your late thirties, your chances of conceiving within a year remain above 50/50. In short, there is no magic cutoff. Age 35 is not a cliff's edge; it's just another step along the runway.

---

9   By the way, if calling 35-year-olds "geriatric" doesn't faze you, then you are too young to have kids.

On First Encounter   After Reframing

I confess that there's not always a "right" framing. In Kahneman's disease scenario, it's accurate to say either "200 people will live" or "400 people will die." The decision is as hard as granite, and any framing that makes it feel easy is a delusion by another name.

Even a deft framing may not yield an immediate answer. For example, when considering a number in Arpeggios, how much runway should you be willing to spend? Seven spots? Ten spots? The unsatisfying truth is that it depends—on the length of your list, the length of your opponent's, and how much runway you each have left. If that's true for a dice game, just imagine the complexity of real-life decisions such as how to treat a pandemic, or when to have a baby.[10]

The best we can do is seek clear and intelligent framings, ones that bring the risks and rewards to light, resisting easy binaries and highlighting authentic trade-offs. Even in a world full of luck, it's incumbent on us to make the best decisions we can.

VARIATIONS AND RELATED GAMES

MULTIPLAYER ARPEGGIOS: You can play with as many as six players. Simply alternate ascending and descending as you go around the circle. The play and passed dice move to the left.

ASCENDER (FOR ONE PLAYER): Begin with a list of **10 blanks**. Roll the dice. Each time you roll, you must **write the number**, in either digit order, somewhere on your list. It need not go at the top; any spot is allowed. The trick is that **your list must only ascend**, never descend. If you fill all 10 blanks, you win. If you run into a number that you cannot place, you lose.

---

10 Easier questions include "how to treat a baby" (gently) and "when to have a pandemic" (never).

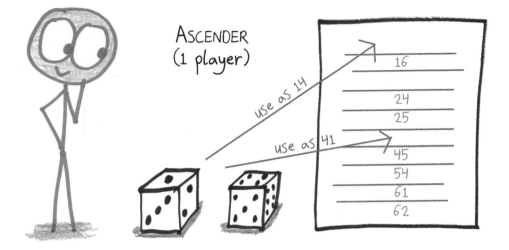

ASCENDER
(1 player)

ASCENDERS (FOR 2 TO 10 PLAYERS): Joe Kisenwether devised this neat little game. Each player begins with a list of **15 blanks**. Every roll of the dice is communal, so after every roll, **everyone must write the number (in either digit order) somewhere on their personal list.** You may not skip a roll. At the end, the person with the **longest unbroken ascending streak wins**. Break ties with the second-longest streak.

streak of 8   streak of 12

OUTRANGEOUS

AN UNCERTAIN TRIVIA GAME FOR AN UNCERTAIN WORLD

I enjoy trivia games: the camaraderie, the tension, the chips, the salsa . . . all of it, really, except the pesky part where I need to know things.

Outrangeous is a game for folks like me. You answer each question ("How many apostles did Jesus have?") not with a specific number, but with a range. Miss the truth (e.g., "50 to 100"), and you score no points (hence, "out-range-ous"). Capture the truth, and you score more points based on how narrow your range is (so "10 to 13" beats "11 to 18").

In the end, the game isn't about how much you know. It's about recognizing what you don't.

HOW TO PLAY

What do you need? Four to eight players (though you can make do with three.) Also, pencils, paper, and—at least for the first few minutes—the internet.

Before beginning, have everybody take five minutes to come up with a few trivia questions whose answers are (a) numbers and (b) easily googled.

So, in this game we just sit around on our phones?

Aww, darn.

Only at the start.

(How old was the oldest-ever gorilla?) (How many sheets are in 1 kilogram of paper?) (How may Australian cities have over 1 million people?)

What's the goal? Each answer is a number. You'll guess a range of values, trying to make it as narrow as possible while still including the true answer.

What are the rules?

1. One player—the judge for the round—announces the trivia question. The other players act as guessers, **each secretly writing down a range of values**.

2. When everyone has committed their answer to paper, the guesses are revealed. The goal is to **capture the true value**, while keeping your range as **narrow as possible**.

3. The judge reveals the true answer. **Anyone who missed the answer—no matter how painfully close they came—receives 0 points**. Instead, the **judge receives 1 point per wrong guess**, as a reward.

4. Then, among the players with the correct answer, **order them from narrowest range** (i.e., most impressive guess) **to widest range** (i.e., least impressive guess.)

5. These players receive **1 point per guesser that they beat**. Note that they all beat anyone who missed the answer.

6. Play enough rounds so that **each person has an equal number of turns as judge**. In the end, whoever has the most points is the winner.

TASTING NOTES

Upon first formulating your guess, you'll feel pretty good.

Then, when the answer is revealed, you'll be shocked how often you missed.

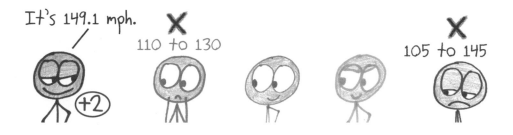

This creates an incentive to "go wide." You can often beat the wrong guessers, and thereby rack up points, merely by admitting your own ignorance.

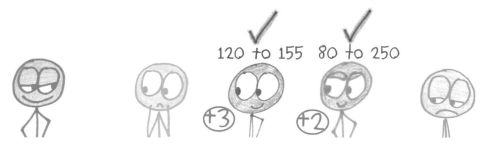

Then again, if *everyone* is going wide, there's an incentive to go a bit narrower. In a world full of people guessing 0 to 1 million, the person guessing 5 to 500 is king. To explore this dynamic, let's focus on a simple question for two players.

We'll roll a 10-sided die. The question is "What number will come up?"

Now, if I guess a wide range, like 1 to 8, then your best bet is to undercut me with 1 to 7. That way, if the answer is 1, 2, 3, 4, 5, 6, or 7, then you win by virtue of your narrower range. Meanwhile, if the answer is 9 or 10, nobody is right, and it's a wash. You'll only lose if the die turns up an 8.

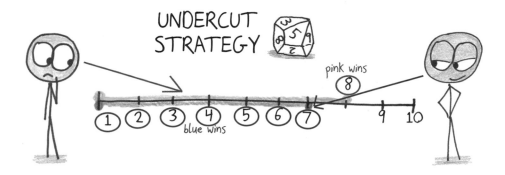

What if I guess a narrow range, like 1 to 3? In that case, your best move is to go as wide as possible: 1 to 10. You'll lose if the die comes up 1, 2, or 3, but you'll win if it comes up 4, 5, 6, 7, 8, 9, or 10. It's worth the trade-off (and better than going for the 1-to-2 undercut).

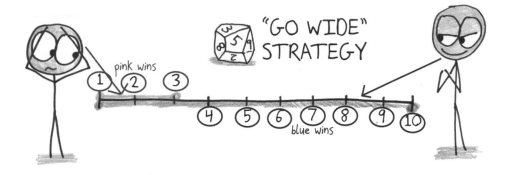

In short: If I go wide, you should go a little narrower, and if I go narrow, you should go wide.

For exactly this reason, I'd be a fool to tell you what range I'm going to pick. Instead, I'm going to randomize my answers. You'd be wise to do the same. With game theory, we can calculate the optimal probabilities:

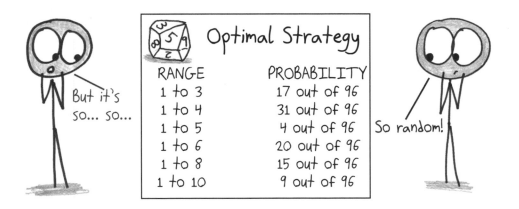

RANGE	PROBABILITY
1 to 3	17 out of 96
1 to 4	31 out of 96
1 to 5	4 out of 96
1 to 6	20 out of 96
1 to 8	15 out of 96
1 to 10	9 out of 96

Weird, right?

Your actual best strategy will vary depending on the question, the score, your own knowledge, and the number of players (the more there are, the wider you want to go). But I hope this gives a taste of the game's subtle pressures.

WHERE IT COMES FROM

In Douglas Hubbard's *How to Measure Anything*, I came across 10 Outrangeous-style questions, along with an instruction: *Make each range wide enough that you're 90% confident of capturing the true answer.* That's 90% exactly: no more, no less. As a math teacher, a probability aficionado, and (as my siblings describe me) "a robot," I felt positive I'd nail it. I'd be 90% accurate, missing one of the 10. Maybe zero or two, depending on my luck.

Instead, I missed four.

Watching my surefire A- pale into a D- prompted a small crisis of confidence. As it should have, because my confidence was the whole problem. My confidence had slipped its leash and was now running amok, barking at squirrels, chasing traffic. How could I trust myself to calculate life's risks and rewards knowing I had such an inflated sense of my own powers?

Inspired and chastened, I developed Outrangeous as a classroom game.[11] Other folks have independently developed the same concept.

WHY IT MATTERS

Because to take calculated risks, you must know the limits of your own calculations.

Humans are not perfect. Your view of the world, just like mine, is a simmering mix of fact and fiction, history and myth, "tomato is a fruit" and "who am I kidding; you can't put tomato in a fruit salad." The question isn't whether my beliefs are true or false. I have true *and* false beliefs, both in abundance. The question is whether I can tell the two apart, and the sorry reality is most of us can't. We carry all of our opinions, right and wrong alike, with a swashbuckling, wholly unearned confidence.

In a classic study, psychologists Pauline Adams and Joe Adams quizzed subjects on how to spell some tricky words, and asked them to rate their confidence in each. Occasionally, folks would say "100%." That means deadlock certainty. Total guarantee. If I created a YouTube supercut of every time you've ever claimed 100% certainty, it should include exactly zero cases in which you were wrong.

Instead, on such 100% answers, the study found a 20% rate of error. "I'm absolutely positive and would bet my cat's life on it" translates to "Eh, call it four out of five."

---

11 I called it Humility at first, because that's what you need to win (and what my exuberant students often lacked). My friend Adam Bildersee later suggested the wittier Outrangeous.

A little overconfidence isn't a crime, at least not in most jurisdictions. It can even help, by giving us the courage to start an ambitious yet likely-to-fail project, such as writing a novel, running for political office, or reaching in-box zero. Still, whenever humans work together, we need to share our knowledge. That's a doomed endeavor if nobody can distinguish their knowledge from their ignorance. What's the point of pooling our money if we can't tell the real bills from the counterfeits?

Luckily, a noble few have learned to navigate these dark tunnels of uncertainty. They are called statisticians, and they will tell you, in no uncertain terms, that nothing is truly certain.

Imagine a study that finds the average American thinks about cheese 14.2 times per day. No matter how careful the researchers, or how tantalizing the Gruyère, there remains a modicum of doubt. Perhaps the true answer is a little lower (because we polled an unusually cheese-loving sample) or a little higher (because our subjects were unusually cheese-averse).

The solution is a confidence interval. Or better yet, a collection of them.

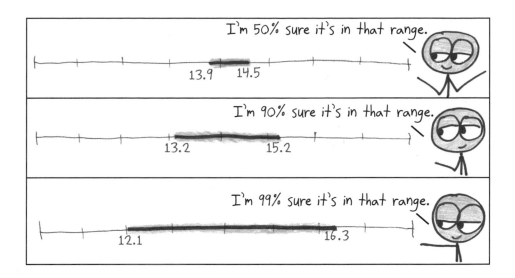

Such intervals embody an inherent trade-off. You can give a narrow, precise range. Or you can give a wide range that's almost certain to capture the truth. But you can't do both at once.

The tighter the range, the greater the risk of missing the mark.

Outrageous demands the same trade-off. You can give a narrow range, which might garner a lot of points. Or you can give a wide range, upping your chances to score at least *some* points. You just can't do both at once.

To execute either strategy, you need to pursue a rarefied psychological state: *good calibration*. This means that your confidence matches your accuracy. When you feel 90% confident, you're right 90% of the time. When you feel 50% confident, you're right 50% of the time. You say what you mean and mean what you say. Subjective feeling aligns with objective success.

To be clear, good calibration is a narrow virtue. If you're 50% confident that sharks are fish (true) and 50% confident that prairie dogs are fish (not so much), then you're well calibrated, but a fool. Meanwhile, if you're 5% confident that testing your bomb will extinguish life on earth, yet you shrug and start the countdown, then you may or may not be well calibrated, but you're definitely a monster.

Good calibration isn't sufficient for good judgment. But it may very well be necessary. Games like Outrangeous offer a unique window into your calibration and a training ground for improving it.

When my wife was in grad school for math, we'd team up with some friends from her program to play bar trivia on Thursday nights. Our team won every week, and usually in the same way: by hanging close in the themed rounds (sports, geography, music, etc.), then surging to victory in the final general knowledge round.

This posed a bit of a mystery. If another team outperformed us on, say, history, science, and film, then shouldn't they beat us on general knowledge, too?

I eventually developed a theory of our strange success. During themed rounds, a team can hand their answer sheet to the relevant specialist—the sports fan, the music expert, the geography whiz—and defer to them. But in general knowledge, no expert reigns. Everyone weighs in. You'll soon have four or five suggestions, one of them probably right. How do you know which? How can the group settle on the correct answer, rather than the most overconfident?

This is where the mathematicians shone. Mathematical research forces you to distinguish carefully between airtight knowledge, credible belief, plausible hunch, and blind guesswork. Our teammates never fought for an answer just because it was their own. Instead, the truth would rise to the top.

The mathematicians were well calibrated.

That's my belief, anyway. It's also possible that by the final round, everyone else was drunk, while the mathematicians held their liquor better. As with anything, I'll never be 100% sure.

VARIATIONS AND RELATED GAMES

RATIO SCORING: Say we're guessing the distance to the moon. I put "3,000 miles to 300,000 miles," while you put "100,000 to 400,000 miles." We both get it right (the truth being 239,000). And, per the rules, my range is a bit narrower. But was mine really the better guess? My lower bound suggests that the moon and Earth might be closer together than New York and London. Your guess seems far more sensible. Shouldn't it score more highly?

The solution: *divide* rather than *subtract*. That is, calculate a ratio, rather than a width. Here, my ratio is 100 (that's 300,000 divided by 3,000), while yours is just 4 (from 400,000 divided by 100,000). Your guess is far more precise.

I recommend this scoring system for **questions where ranges may span several orders of magnitude** (e.g., "number of slot machines in Las Vegas"). For more restricted ranges (e.g., the age of a particular celebrity), the original scoring system works fine.

THE KNOW-NOTHING TRIVIA GAME: Years ago, in the course of a long plane flight, the mathematician Jim Propp and two friends invented this strange jewel of a game. It's almost a contradiction in terms: a trivia game you can play without ever finding out the answers.

It works for any odd number of players. Take turns coming up with a **numerical trivia question** (e.g., "How many home runs did Barry Bonds hit in his career?"); then all of you (including the question asker) **write down a secret guess**. When the guesses are revealed, **the winner is whoever's guess is in the middle**.

For example, if the three guesses are 900, 790, and 2,000, then the person who guessed 900 is the winner. Never mind that the truth is 762. You're not trying to guess the *right* answer, but the answer that will land *between* your friends' answers (though in practice that usually means just giving it your best guess).

ADVICE FOR WRITING QUESTIONS

Spend 10 minutes on Google and/or Wikipedia before the game begins, so that by the time it's your turn to judge, you've got two or three questions ready to roll.

Play to your audience. Absurdly hard questions are no good; everyone just shrugs and gives a very wide range. The best questions are tantalizing: you don't know the answer, but feel like you should.

Phrase questions as precisely as possible. Where relevant, specify units ("distance *in miles*"), dates ("population *as of 2019*"), and sources ("the film's budget *according to Wikipedia*").

Here are some suggestions. You can also use these to inspire other ideas—just swap in a different celebrity/place/world record/piece of pop culture.

- Age of Jamie Foxx
- Age at which Abraham Lincoln died
- Age of the oldest-ever manatee
- Amount of money Judge Judy makes per year
- Current day of the month (without looking)
- Distance to the moon in miles
- Distance from NYC to LA (as crow flies)
- Height of the tallest-ever ice cream cone
- Height of the tallest-ever WNBA player
- Hottest land temperature ever recorded
- Length of "Bohemian Rhapsody"
- Length of Canada's coastline
- Length of every *Simpsons* episode ever, if watched back to back to back
- Length of Nelson Mandela's prison term
- Length of the longest fingernails ever
- NBA season record for rebounds per game
- NFL single-season record for most interceptions thrown
- Number of episodes of *Sesame Street*
- Number of in-ground pools in Texas
- Number of lakes in Minnesota
- Number of goldfish crackers (out of 10) that I will successfully toss into this bowl from a distance of six feet
- Number of novels by Agatha Christie
- Number of species of penguin
- Number of bird species that can fly backward
- Number of studio albums by Jennifer Lopez
- Number of US states with wild alligators
- Number of words in Hamlet's "to be or not to be" soliloquy
- Percentage of the presidential vote won by Ross Perot in 1992
- Percentage of US adults that believe chocolate milk comes from brown cows
- Percentage of the US that identifies as male
- Population of Atlantic puffins worldwide
- Population of South America

- Pounds of trash generated per day by the average US citizen
- Price for which the most recent Van Gogh painting sold
- Publication date of first Harry Potter book
- The 1,000th prime number
- Time it will take this ball I'm holding to stop rolling when dropped from waist height
- Time it would take to drive from here to the Empire State Building, per Google Maps

- Total box office for *Avengers: Endgame*
- Total value of the Disney Corporation
- Weight of an average humpback whale
- Year in which the last French king was born
- Year in which first Nobel prizes were given

PAPER BOXING

A GAME OF NARROW VICTORIES

This pencil-and-paper boxing simulation is all too easy to confuse with the real sport. Both pit two foes in 15 rounds of merciless combat, both can be played in shiny shorts, and both go best with someone in your corner mopping your sweat and muttering, "You got this, champ, you got this." Still, if you look closely, you'll spot a few telltale differences.

Regular Boxing	Paper Boxing
Multimillion-dollar prizes	Zeromillion-dollar prizes
World champion compelled to wear awkwardly heavy belt	World champion allowed to wear any belt whatsoever
Extremely dangerous to face Floyd Mayweather	Only somewhat dangerous to face Floyd Mayweather
Gladiatorial appeal to our primal competitive instincts	Gladiatorial appeal to our . . . wait, this isn't a difference

Anyway, please don't punch your Paper Boxing opponent in the face. The numbers will handle that for you.

HOW TO PLAY

What do you need? Two players, two pens, and four sheets of paper. Set two sheets aside. With the remaining two, each player draws a 4-by-4 grid, leaving blank the upper-left corner, and **secretly filling the other cells with the numbers 1 through 15**.

What's the goal? Win a majority of the match's 15 rounds.

What are the rules?

1. Sit side by side, **revealing your grids** to one another. These shall remain faceup and visible for the whole game. Then, take your other sheet of paper and **secretly write down your first number**. It must be one of the three numbers adjacent to your blank "starting" square.

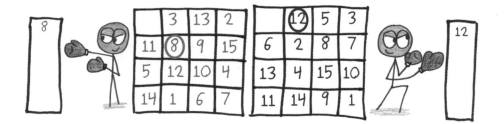

2. Reveal your choices to each other. **Whoever chose the higher number wins the round**, scoring 1 point. In case of a tie, neither player scores. Either way, each player now **records their path** by drawing a line from their blank square to their chosen square.

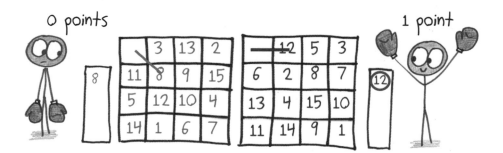

3. Repeat this process, moving from your most recent number to any of its neighbors. Your path may **cross itself along diagonals**, but **cannot revisit any squares**. In every round, **the higher number scores 1 point**.

4. If you manage to **trap yourself**, so that your path cannot continue, then you have in effect chosen the number **0 for all remaining rounds**.

5.   The champion is **whoever wins more rounds** (i.e., scores more points).
     Ties are possible.

TASTING NOTES

No way around it: You're going to lose a few rounds.[12] You should view those losses
as opportunities. In your defeats, try to burn weak numbers (like 1, 2, and 3) while
your opponent squanders strong ones (like 13, 14, and 15). The more resources
they waste on lopsided victories, the more razor-thin wins you can eke out.

     The strategy, in a nutshell: *Lose big, win small.*

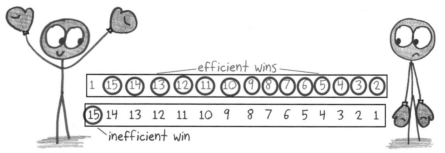

     After that, there's a second strategic layer: charting your overall path.
There are myriad ways to navigate the board. Almost 38,000 paths visit
all 15 squares, and another 300,000 visit the majority before getting stuck
somewhere.[13] Poor choices early on may foreclose future options, leaving your
final moves predestined and your opponent in total control. But with careful
planning, you can keep your options open until the very end.

---

12   Unless you tie every round, but that's neither fun nor likely.
13   91,000 paths miss one square, and another 102,000 miss two squares. But there are some stinkers: 22
paths end after just five moves.

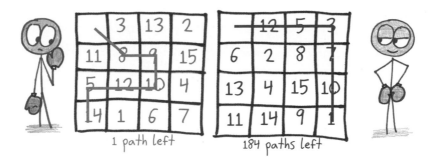

1 path left   184 paths left

The third (and, to me, most mystifying) layer of strategy is designing your board in the first place. There are more than a trillion ways to arrange the numbers 1 to 15, and I find it almost impossible to tell the good from the bad from the ugly. It's like trying to sift, from a barrel full of dried beans, the handful of cow-worthy magic ones.

My advice: Scatter your numbers more or less randomly. At least that'll keep your options open.

WHERE IT COMES FROM

Sid Sackson introduced Paper Boxing in his 1969 book *A Gamut of Games*. I toyed with various modifications to his clever original, and implemented two. First, a minor change: If you trap yourself, you do not immediately lose (as Sackson proposed) but instead score zero for the remaining rounds. As someone who occasionally walks into parking meters, I wanted to relax the penalty for a wrong turn.

Second, a major change. In Sackson's original, numbers are chosen in full view, one player at a time, starting with whoever won the last round. The second mover can see exactly what the first mover picked. I prefer simultaneous secret selection, and not just for the alliteration: To my mind, it gives the game more punch.

WHY IT MATTERS

Because it's threatening US democracy.

Let me back up. *Lose big, win small* isn't just about Paper Boxing. It applies anytime two conditions are met: (1) You need to spread finite resources across multiple efforts, and (2) there's a sharp cutoff between success and failure. For example, a coldblooded, grade-maximizing student would rather earn an A with a bare-minimum 93% than a comfortable 99%. Those six extra points come at the expense of time for another class, or for Call of Duty. A big win isn't a win at all. It's a misfire, burning resources you could have spent elsewhere.

I don't mean to say that grade-maximizing students threaten our civic life. Maybe they do, but the bigger threat is our newfound skill at a high-stakes legislative version of Paper Boxing: the risk-and-reward game known as *gerrymandering*.

Here's how you play. In a given state, roughly half of the people vote for red elephants, and half for blue donkeys. Your task is to carve up the state into equal-sized districts. In each, the political party with more voters scores 1 point (i.e., one seat in the legislature). Can you maximize your team's score?

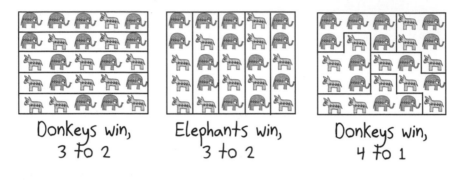

Voters are a finite resource, spread across multiple districts, in each of which there's a sharp line at 50%. Thus, you want to lose big in a few districts, so you can win narrowly in the rest.

Think in terms of wasted votes. It's delightful to win 1,001 to 1,000: You didn't waste a single vote, and your opponent wasted all of theirs. Just as delightful, though, is *losing* 2,001 to 500: Sure, you wasted 500 votes on the loss, but your opponent wasted 1,500 votes on a needlessly wide margin of victory. By tailoring the map to waste your opponent's votes, you can lose the overall tally and still come out ahead.

The US has been playing this game for centuries. The word "gerrymander" was coined in 1812 by the *Boston Gazette*, as a portmanteau of "Gerry" (as in Massachusetts governor Elbridge Gerry) and "salamander" (because Gerry proposed districts so twisted that they looked like writhing lizards).[14]

---

14 My friend David Litt, author of *Democracy in One Book or Less*, makes a crucial point here: "It appears the editors of the *Boston Gazette* had absolutely no clue what a salamander looks like." David explains: "The creature depicted in the famous cartoon had the wings of a dragon, the talons of an eagle, the neck of a python, the beak of a vulture, and the teeth of a piranha . . . The entire image was a crime against herpetology."

For most of our history, gerrymandering was more of a nuisance than a threat. It's hard to draw districts by hand, and besides, a minor shift in the political winds can turn narrow victories into narrow defeats. Held back by caution and miscalculation, gerrymandering was not so much a science as a pseudoscience, the phrenology of power grabs. By 2000, aspiring gerrymanderers were still stumbling in the dark.

Then came Big Data. By 2010, parties could simulate election results for billions of maps and select from these the most ruthlessly efficient. And so they did. In 2018, Democrats earned 53% of the vote in Wisconsin, yet thanks to an overpowering gerrymander, Republicans won 63% of the State Assembly districts.

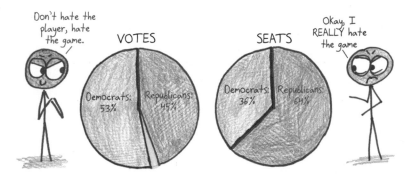

Why not ban salamander-shaped districts? Well, as mathematician Moon Duchin points out, (a) it's tough to define "salamander-shaped," and (b) even if you could, it wouldn't solve the problem. There'd still be quintillions of biased maps to choose from. You can build an overwhelming advantage out of districts with perfectly nice shapes.

Okay, so why not require the proportion of seats to match the proportion of votes? Well, even without gerrymandering, that's not what happens. Try mixing red and blue paint in a 55/45 ratio, then dividing the mix into smaller batches. The red paint will "win" every batch. Some countries have proportional systems, but ours, for better or worse, has never been one.

So what can we do?

"In representative democracy," Moon Duchin once told *Quanta* magazine, "you have a lot of different ideals in tension." Majority rule. Minority voice. Districts that correspond to actual communities rather than arbitrary geometry. "It's about understanding how your priorities trade off rather than trying to find the 'best' of something."

Games have winners and losers. But at their best, democracies don't. They have conflicts, which lead to conversations, which lead to compromises. Even if we never get to consensus, at least we all live to play another day.

VARIATIONS AND RELATED GAMES

PAPER BOXING CLASSIC: Instead of picking numbers in secret, pick them in public, one player at a time, starting with whoever won the last round. On the first turn of the game, start with whoever has the largest sum of numbers touching the blank square.

PAPER MIXED MARTIAL ARTS: Instead of placing the numbers 1 to 15, you may **fill your board with any set of 15 whole numbers** (including 0) **whose sum is 120**.

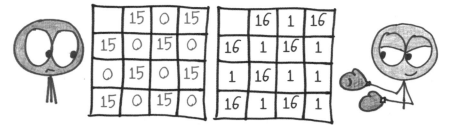

If you wish, you may somewhat limit the variety of possible boards, with rules like these:

- You must include a number that is 30 or higher.
- You cannot use any numbers higher than 15.
- You may use at most five different numbers.

- You must use at least 10 different numbers.
- One player must use all even numbers; the other must use all odd (summing to 121).

BLOTTO: This lightning-quick game, a kind of simplified Paper Boxing, played a key role in the development of game theory. Each player writes a **secret list of three whole numbers in order from smallest to largest.** Repetition is allowed, but the numbers must **add up to exactly 20.** Then, compare lists. At each spot, the **largest number wins,** and whoever has more winning numbers wins the game.

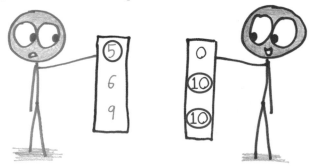

For a multiplayer game (or just a more involved one) you can tweak the parameters. For example, write five numbers summing to 100, or 10 numbers summing to 500.

Blotto is traditionally presented as a military game: Each player has 20 "troops," deployed across three "battlefields." But my preferred metaphor comes from math teacher/golf coach Zach McArthur, who compared the game to "a par 4 hole that has a bunker right on the peak of the dog-leg." He elaborated: "You start out way away and each successive time you play the hole you try to shave the corner closer until you end up in the bunker and then start out way wide again." I have no idea what he's talking about, and I know exactly what he means.

FOOTSTEPS: In this two-player game, you vie for the affections of a lovable donkey. The donkey begins in the middle of a field, three steps from each opponent. Each player begins with a **bag of 50 oats**, and on each turn, **secretly chooses a number of oats to offer**. The donkey then **takes a step toward the larger offer**. Oats are never regained, even if the donkey rejects your offer. **You win if the donkey reaches you.**

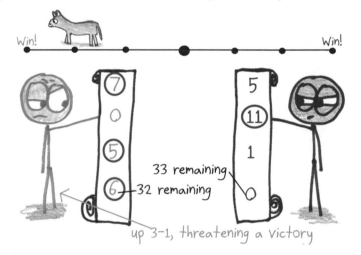

If both players run out of oats, then the **winner is whoever the donkey is closer to**. But watch out: If you spend all of your oats while I still have some, then you can only watch helplessly as I win by a score of 1 to 0, over and over again, until the donkey's love is mine.

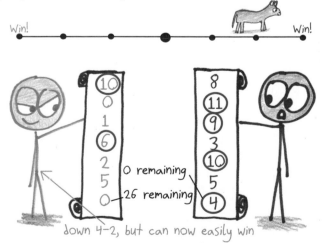

RACETRACK

A GAME OF PHYSICS

I believe there's only one game with these two items side by side on its résumé:

1. Beloved by bored students, who play it discreetly to pass the time in class.

2. Beloved by science teachers, who use it to explain concepts like inertia and acceleration.

Perhaps somewhere, on some magical day, a pair of goof-off high schoolers unwittingly passed a physics lesson playing the very same game they were trying to ignore.

HOW TO PLAY

What do you need? Two players, two colored pencils, and a sheet of graph paper. Using a thick black pen, **draw a racetrack**. It can be as wobbly and curvy as you like, as long as it's clear which grid intersections lie inside the track, and which ones lie outside. Then, **draw a start/finish line**, and **mark a starting point for each car**.

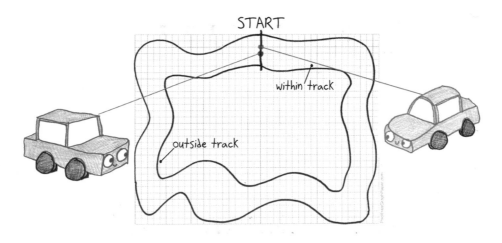

What's the goal? Harness your inertia to cross the finish line before your opponent.

How do you play?

1. On each turn, move your car **a certain distance in a straight line**. (I'll explain the details under rule 4.) **You cannot move to the same point where your rival currently is**, though you can move to a point where they have already been.

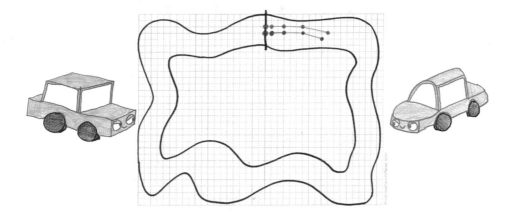

2. **If you hit a wall or exit the track, you lose two turns**. Afterward, you begin anew from the point on the track nearest to where you exited.

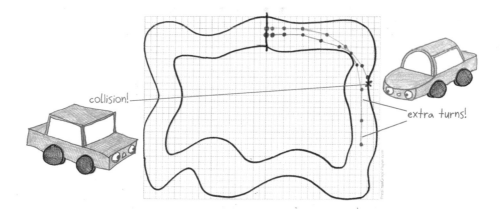

collision!

extra turns!

3. First to complete the course and **cross the finish line is the winner**. If both players complete the course on the same turn, farthest past the finish line wins.

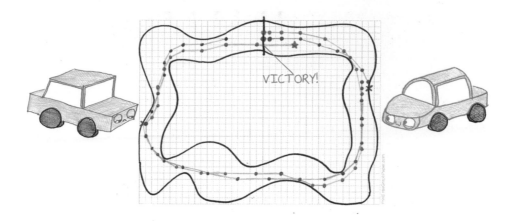

4. Now, the most essential rule of all: How, exactly, do the cars move? It's up to *inertia*. **Whatever your car did on your last turn, it will do again on this one**—except that you can, if you like, **adjust its motion one unit in a vertical direction and one unit in a horizontal direction**.

Let's say your last turn brought you **3 steps to the right**. This time around, you have three options for horizontal motion: **2, 3, or 4 steps right**.

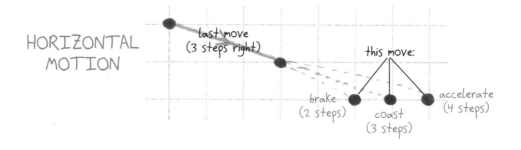

And if your last move brought you **1 step down**, then you have three options for vertical motion: **0, 1, or 2 steps down**.

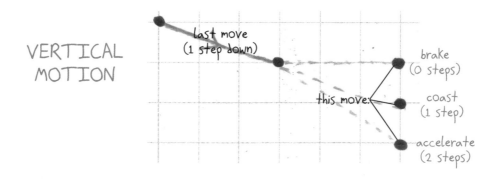

These choices happen independently. For example, you can accelerate horizontally while braking vertically, or vice versa. That yields **nine possible moves in all**.

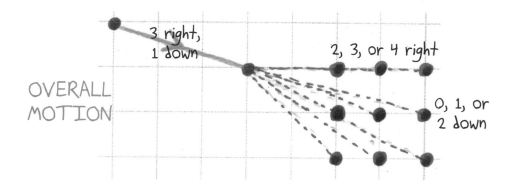

Note: When starting the game (or restarting after a crash), assume that your "last turn" consisted of 0 steps in any direction.

TASTING NOTES

With its agonizing wrecks and dramatic escapes, Racetrack is perhaps the most exciting fate that can befall a piece of graph paper.

Better yet, once you master the rules (which may take a little patience), the game will begin to evoke real-life physics. Why can't you move 4 squares to the right, then 3 squares to the left? Same reason you can't pull an instantaneous U-turn at freeway speeds.

You'll learn to calibrate your speed carefully. Go too slow on a straightaway, and you waste precious time. Go too fast, and you'll struggle to brake for the next corner.

In short, it's just like real car racing, but with fewer carbon emissions.

WHERE IT COMES FROM

As with many folk games, Racetrack's origins are lost to time, though "Western Europe in the 1960s" seems like a credible guess. In 1971, a French version was published under the devastatingly cool name Le Zip. In a 1973 *Scientific American* column, Martin Gardner (having learned the game from a computer scientist who picked it up in Switzerland) described it as "virtually unknown" in the US.

Anyway, once it hit the US in 1973, it became a schoolhouse favorite. A rudimentary computer version at the University of Illinois proved such a popular time waster that school authorities banned it for a week.

WHY IT MATTERS

Because risks and rewards lurk everywhere—even in the cold austerity of a deterministic world.

"Every body," wrote Sir Isaac Newton, "perseveres in its state of rest, or of uniform motion in a right line, unless it is compelled to change that state by forces impress'd thereon." For centuries, we've been paraphrasing Newton's wisdom to ourselves, with phrases like "Every body at rest tends to stay at rest," "Every body in motion tends to remain in motion," and "Everybody, rock your body right. Backstreet's back, all right."

All of these formulations boil down to the same thrilling and terrifying idea. Orbiting planets, plummeting apples, resurgent 1990s boy bands—they all obey a few universal laws of motion.

They keep doing their thing, until something tells them otherwise.

Racetrack puts this principle into action. Your car moves just as it did last time, subject to a little modification from you. It's a pencil-and-paper version of inertia, so elegantly designed that *Car and Driver* magazine hailed its verisimilitude as "almost supernatural."

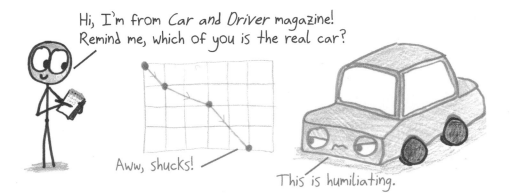

Though no randomness is involved, Racetrack provides an elemental experience in risk and reward. On the one hand, you can play it safe: moving just a few squares at a time, keeping away from the walls, and otherwise minimizing the danger of a crash. On the other hand, you can go for broke: accelerating to high speeds, cutting close to the walls, and crossing your fingers that you avoid a spectacular wreck.

As in physical car racing, the reward is speed, the risk is disaster, and it's your job to strike a balance between them.

The other risk and reward games we've explored each contained some unknowable element. You cannot foresee how the dice will land in Arpeggios, or what your opponent will choose in Undercut, Outrangeous, or Paper Boxing. Racetrack, by contrast, is pretty much all knowable. You could, in theory, calculate the optimal route for your car, planning out your path to the very last square, before the game even starts. It's just that you won't, because it'd take you ages, and your opponent would wander off to make more interesting friends.

The risks and rewards here come not from the unknowable, but merely from the unknown.

To some philosophical types, this distinction matters a lot. Is the future fundamentally unknowable (because it could still unfold in multiple ways)? Or is it merely unknown (because it's beyond our power to calculate)? The first view allows for free will; the second might not. Seems like a big deal.

Then again, what's the difference between a future I *can't* foresee, and a future I *don't* foresee? Either way, the best I can do is to compare the possible outcomes, weigh their probabilities, and take a measured risk.

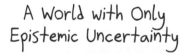

A World with Genuine Ontological Uncertainty

A World with Only Epistemic Uncertainty

CAN YOU SPOT 10 DIFFERENCES?

(answer: no, because the subjective experience is identical)

Racetrack offers limited choices from turn to turn. When speeding like a demon, you can (a) make your speed a little *more* demonic, (b) make your speed a little *less* demonic, or (c) keep it *equally* demonic. Not much control there. Yet, over a long enough time horizon, you can do anything: speed up, slow down, reverse directions, trace out a lazy figure 8 . . .

In Racetrack, as in life, inertia poses no obstacle that patience and will cannot overcome.

In the end, Racetrack's simultaneous appeal to teachers and students shouldn't surprise us. It is a simple, rule-governed version of reality, with just enough free choice to make things exciting. One name for this is "mathematical model"; another, just as valid, is "game."

VARIATIONS AND RELATED GAMES

Racetrack, like Paper Boxing, lends itself to all sorts of tasty twists and house rules. I offer just a small sampling.

CRASH PENALTIES: You can increase the penalty for crashing—for example, you miss three turns. Or, as in Martin Gardner's version of the rules, you may even suffer an instant defeat.

MULTIPLAYER RACETRACK: The game works with three or even four players (though you may want to combine two sheets of paper for a longer and wider track).

OIL SPILLS: Shade in a region to mark it as "slippery." Cars passing through this region may not accelerate or decelerate at all; they must continue in the same direction at the same speed.

POINT GRAB: Instead of drawing a track, place 20 or so "flags" by marking random intersections across the page. Begin in the corner, and move according to the usual rules. First person to reach a given flag (by ending a turn precisely on top of it, not just passing over or through it) scores a point. Most flags wins.

SLANTED START: To diminish the advantage of the first player, you can draw a slanted starting line, then allow the second player to choose which starting point they want.

THROUGH THE GATES: Instead of drawing a track, place a sequence of numbered "gates" (1, 2, 3, 4 . . .) around the graph, each two or three squares wide. The winner is the first to pass through all of the gates in numerical order.

A QUICK SHUFFLE OF RISK-AND-REWARD GAMES

Here are six rapid-fire games. Each is a tiny model universe, with its own juicy trade-offs and vexing decisions, a kind of dress rehearsal for negotiating the risks and rewards of real life.

I don't mean to promise that dice games can teach you investment strategies, or that a few rounds of rock-paper-scissors will improve your negotiating tactics. Games are simple; life is complex. Games have known odds; life has unknown oddities. Still, just as a stick-figure drawing captures something essential about the human shape, I believe these simple games capture some real truth about our messy world.

PIG

A DICE GAME OF PRESSING YOUR LUCK

Many games ask you to press your luck. Think of classics like blackjack (take another card, or quit before you bust?), *Wheel of Fortune* (spin again, or solve the puzzle?), *Who Wants to Be a Millionaire?* (try the next question, or bank your winnings?), and *Deal or No Deal* (deal, or no deal?). Perhaps the simplest of all such games is Pig, for two to eight players. On your turn, **roll a pair of dice as many times as you wish, adding their sum to your score each time, and stopping whenever you want**. The winner is the first to reach 100 points.

23 points? That'll do, Dice. That'll do.

There are some bonuses: **doubles score twice their sum** (e.g., 5 + 5 scores 20), and even better, **snake eyes (i.e., 1 + 1) score 25**.

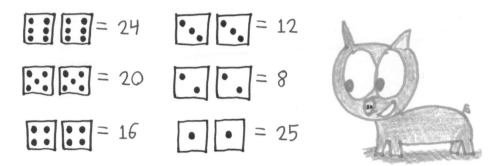

But watch out: if you **roll a 1 plus any other number, then you lose possession of the dice** with 0 points to show for your turn. (Points from earlier turns remain unaffected.) This happens on roughly 28% of rolls.

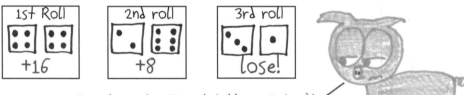

I got greedy. I acted like... I don't know, some kind of gluttonous animal.

Pig offers a dynamic you see in dating, investing, and mountaineering: Do I stop now, or keep going? Settle for what I've got, or risk disaster to seek greater glory? The difference in Pig is that there's a provisional right answer, an optimal way of maximizing your average score per turn. (See the Bibliography for spoilers.)

For a simpler version, play with a single die, each roll scoring its face value, except for 1, which wipes out your score and ends your turn.

Math teacher Katie McDermott also told me about a classroom version. All students begin standing, and the teacher rolls a single pair of dice, which applies to all students. After each roll, each student decides whether to remain standing (and thus risk further rolls) or to sit down (and thus end their participation in the round). The high score after five rounds wins.

CROSSED

A GAME OF SPIDERWEBS

I found this jewel in the teeming pages of Ivan Moscovich's *1000 Playthinks: Puzzles, Paradoxes, Illusions, and Games* (specifically, it's #216). I later created my own physical prototype, played by stretching rubber bands between 16 nails driven into a piece of wood. But all you really need are two players, two colors of pen, and paper.

To begin, draw 16 dots in a square as shown. Then, take turns **connecting two unused dots with a straight line**. The dots cannot be on the same side of the square.

You score **1 point for every time you cross an opponent's line, and 2 points for every time you cross one of your own**. Make sure to keep score as you go.

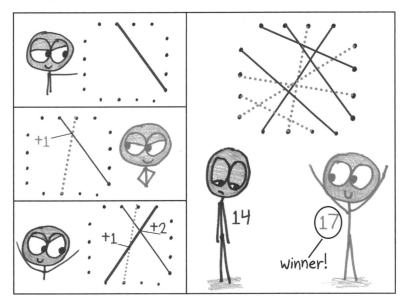

Keep playing until no more moves are possible (either because all dots have been used, or because the only unused dots are on the same side of the square). **The higher score wins.**

Lasting only eight moves, one less than the reviled tic-tac-toe, Crossed may feel simplistic. Yet with dozens of choices per move, its game tree branches rapidly outward, like cracks in shattered glass. The scoring system creates a pleasing tension, a classic trade-off of risk against reward: Short moves deprive you of the chance to self-cross, while long moves leave you vulnerable to an opponent's cross. You're left feeling like a rope pulled from both ends.

ROCK, PAPER, SCISSORS, LIZARD, SPOCK

AN EXPANSION PACK

This game comes from visionaries Karen Bryla and Sam Kass. Frustrated by how often rock-paper-scissors ends in a tie, they added two new gestures: lizard and Spock. This reduces the chances of draws while preserving the game's symmetric structure. Each gesture defeats two others, is defeated by two more, and draws against itself.

To play, just **count to three, and then simultaneously reveal your gesture**. Arrows in the diagram point from winner to loser.

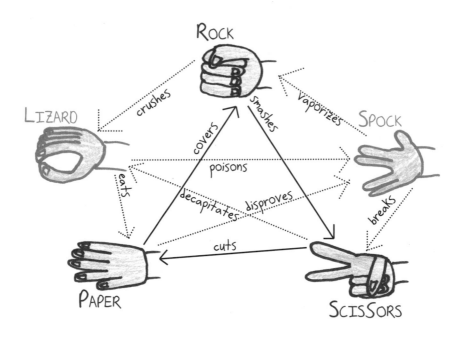

Isn't that a pleasing structure? If A beats B, then you can always find some C that loses to B yet beats A, thereby completing the cycle. No wonder the game earned a giddy shout-out on the hit sitcom *The Big Bang Theory*.

It can be expanded further, to any odd number of gestures: 7, 9, 11, 13, and so on. One dauntless enthusiast spent a year creating a 101-gesture version. Where the original rock-paper-scissors has three rules to memorize, and the lizard-Spock expansion has 10, this ultimate game of gestures has 5,050 rules: "vampire teaches math," "math confuses baby," "baby becomes vampire," and so on. That may be a little much, even for Sheldon.

101 AND YOU'RE DONE

A GAME OF PLACE VALUE

Math educator Marilyn Burns devised this charming game (best with two to four players) to teach place value to elementary schoolers. Finding the optimal strategy is a pleasant, lighthearted puzzle, even for decrepit adults like myself.

Take turns rolling a standard six-sided die. **After each roll, decide whether to leave the number as it is (e.g., 3), or to multiply it by 10 (e.g., 30),** before adding it to your score. Each player should **roll six times** in total. The goal is to **get as close as you can to 100, without going over.** Any value over 100 counts as 0.

Play five rounds; whoever wins the most rounds is the champion.

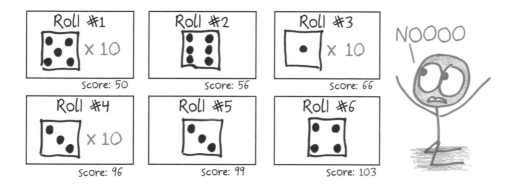

You may feel drawn to the so-called greedy algorithm: always multiply by 10, as long as there's space to do so. That's a fine starting point, but it's foolish to leap up to 96 with two rolls still remaining. Greater caution is required.

To spice things up, you can keep each player's first roll secret; after that, allow everyone to see each roll, but keep private the decision of whether to multiply by 10. Then, after the final roll, reveal all of the choices and results.

THE CON GAME

A REAL-LIFE MASSIVE MULTIPLAYER GAME

Designed by James Ernest, the Con Game can unfold over hours, or even days. It's ideal for a retreat, convention, camping trip, or family reunion. If rock-paper-scissors is like batting practice, then the Con Game is like full-fledged baseball, elaborating a repetitive drill into a sprawling, immersive experience. It also goes well with Cracker Jacks and organ music.

A new player can join at any time and, upon doing so, receives **10 blank cards**. Number them 1 to 10, and on each, write your name, plus either **"rock," "paper," or "scissors."** You can use any ratio that you like. For example, all 10 can be "scissors."

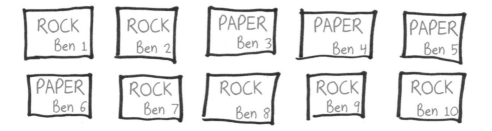

When you come across another player, you may interact in two ways:

1. **Fight.** Each player selects a card in their possession and simultaneously reveals it. The **winner of the fight keeps both cards. If it's a draw (e.g., rock vs. rock), the higher number wins.** If it's *still* a draw, then no cards change hands. You must, if challenged, give an opponent at least one fight. After that, you may decline the challenge until you have each fought someone else.

2. **Trade.** You may **exchange cards, one for one**, with any other willing player. In trading, you may conceal as much information as you wish, but you must not lie.

End the game at a predetermined time (e.g., the start of dinner). Look through your cards for any with your opponents' names on them, and pick out the **highest value from each opponent**. Add these up, and that's your score. (Your own cards are worth nothing.) **Highest score wins**.

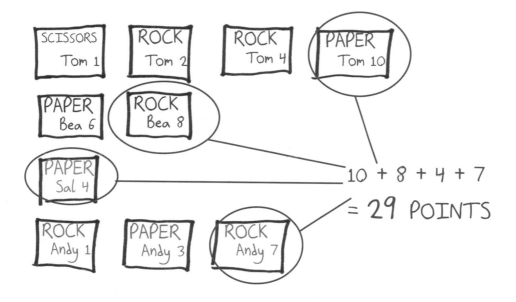

For a really wacky experience, combine the Con Game with Rock-Paper-Scissors-Lizard-Spock.

BREAKING RANK

A TRIVIA GAME

Like Outrangeous, this is a press-your-luck trivia game, where the most important skill is knowing how much (or how little) you know. The goal is simple: Create the longest list that you can, without making any errors.

But fair warning: "Simple" is not a synonym for "easy."

To begin, one player (acting as judge) picks a **group of items**, such as the seven continents, and a **statistic on which to rank them**—say, their land area. I find that pre-specified groups of four to eight items work best, but if you want, you can leave it more open-ended (e.g., the group can be "countries in the world").

Now, each guesser's job is to **list as many continents as they like, in decreasing order of land area.** If your list is correct, with each continent smaller than the last, then **you'll score 1 point per item.**

But **if you make any errors—if, at any point, a larger continent appears below a smaller one—then you'll score nothing,** and the **judge will receive a point** for stumping you.

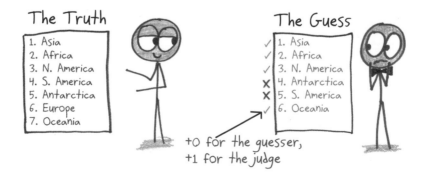

Racking up points *should* be easy. You can always take a guaranteed 1 point, and even a random guess has a 50/50 shot at scoring 2 points. Yet somehow I always push my list one item too far. Heaven help you if you list all seven items: There are 5,040 possible orders, and 5,039 are wrong.

As in Outrangeous, the role of judge rotates, and you should **spend 10 minutes coming up with questions before you start playing**. Make sure to pick well-known items (e.g., celebrities everyone has heard of) and unambiguous statistics (e.g., "number of monthly listeners on Spotify," not "most popular"). Feel free to pick a preexisting category of items (like "countries in Europe") or to handpick a few items (like "France, Germany, Italy, Spain, and the UK"). When in doubt, Wikipedia is a great source.

Finally, a few suggestions for questions:

- **Countries** [e.g., France, Germany, Italy, Spain, the UK] by **population**
- **Continents** [Africa, Antarctica, Asia, Europe, North America, Oceania, South America] by **coastline length**
- **US states** [e.g., Arkansas, California, Kansas, Kentucky, Iowa, Nebraska] by **order of joining the US**
- **Countries** [e.g., Australia, Indonesia, Japan, Mexico, Philippines] by **GDP**
- **Musicians** [e.g., Ariana Grande, Beyoncé, Ed Sheeran, Rihanna] by number of **Instagram followers**
- **Musical artists** [e.g., Coldplay, Kanye West, Queen, Taylor Swift] by **number of studio albums**
- **Songs** [e.g., "Bohemian Rhapsody," "Don't Stop Believin'," "Livin' on a Prayer," "Take on Me"] by **release date**
- **Albums** [e.g., *Abbey Road, Help!, Revolver, Rubber Soul, Sgt. Pepper's Lonely Hearts Club Band*] by **length in minutes**
- **Movies** [e.g., *12 Years a Slave, The King's Speech, The Departed, A Beautiful Mind, Forrest Gump*] by **number of Oscar nominations**
- **TV shows** [e.g., *Everybody Loves Raymond, The Fresh Prince of Bel-Air, Friends, Seinfeld*] by **number of episodes aired**
- **Actors** [e.g., Chris Evans, Chris Hemsworth, Chris Pine, Chris Pratt, Kristen Stewart] by **number of Twitter followers**
- **Books** [e.g., *Beloved, Infinite Jest, 100 Years of Solitude, Slaughterhouse-Five*] by **number of Goodreads reviews**
- **Authors** [e.g., Mark Twain, Charles Dickens, Virginia Woolf, H. G. Wells] by **number of novels**
- **Politicians** [e.g., Al Gore, Hillary Clinton, John Kerry, Howard Dean, Joe Biden] by **age**
- **Planets** [Mercury, Venus, Earth, Mars, Jupiter, Saturn, Uranus, Neptune] by **number of moons**
- **Birds** [e.g., bald eagle, flamingo, grey heron, pelican] by **wingspan**

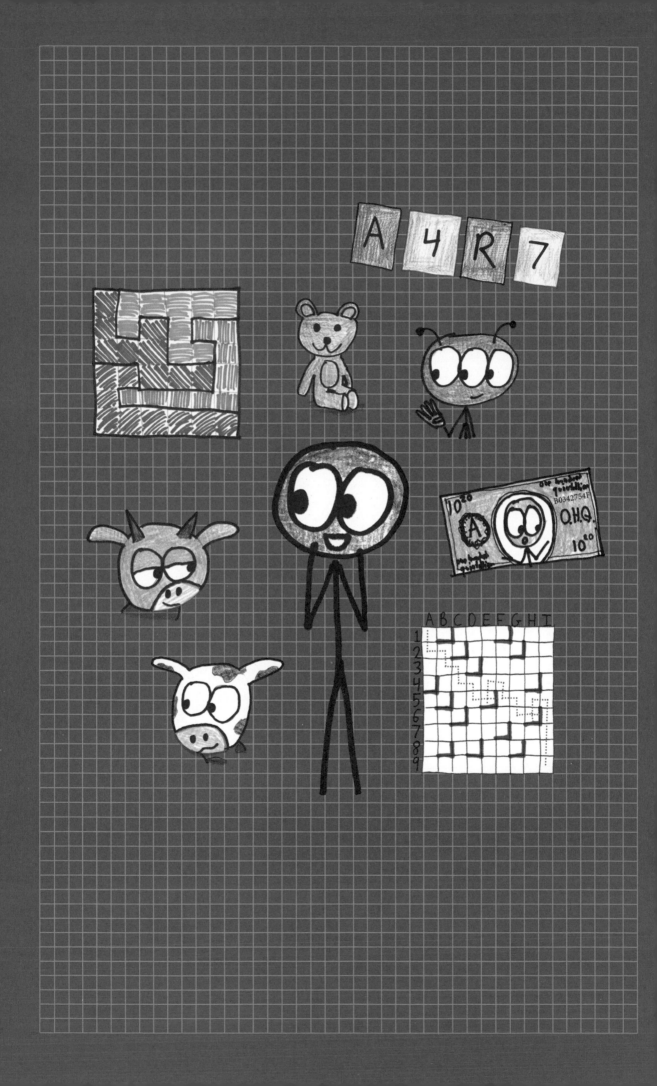

V

INFORMATION GAMES

I GREW UP about a mile from Claude Shannon's house. We never hung out—I blame the 61-year age gap—but I wish we had, because his home was a museum of bizarre inventions. A fleet of unicycles. A flame-throwing trumpet. A juggling robot. A Roman numeral calculator. My favorite is the so-called Ultimate Machine: a box with a switch that, when flipped to On, prompted a disembodied hand to emerge from within, and grouchily flip the switch back to Off. I consider it a kind of existential snooze button.

That said, you can witness Claude's greatest invention just by searching Google for the 1948 paper *A Mathematical Theory of Communication*. It's the paper that revolutionized electronic messaging, the paper that transformed the fuzzy notion of "information" into a precise, measurable quantity.

The paper that birthed information theory.

Just as a measuring cup doesn't care what substance fills it, information theory is indifferent to what the information conveys. "The 'meaning' of a message," Claude explained, "is generally irrelevant." Instead, we imagine each message as being chosen from a list of possibilities. The more possibilities to choose from, the more information conveyed in the choice. A message's information is defined by what you might have said, but didn't.

It's not as crazy as it sounds. If your friend Allegra always says "I'm doing well," no matter how she's feeling, then her utterance conveys zero information. By contrast, if your friend Honestia always tells the truth, then her "I'm doing well" conveys real information, because she chose it over other active possibilities.

The same message may thus convey different amounts of information, depending on context. Saying "I like turtles" when I just asked "Do you like turtles?" doesn't reveal much; I mean, what else were you going to say, that you don't like them? But saying "I like turtles" when I just asked "Tell me about yourself" is *much* more informative. You could have said so many other things.

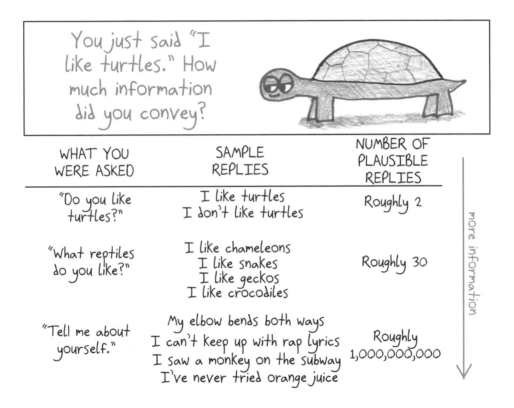

WHAT YOU WERE ASKED	SAMPLE REPLIES	NUMBER OF PLAUSIBLE REPLIES
"Do you like turtles?"	I like turtles I don't like turtles	Roughly 2
"What reptiles do you like?"	I like chameleons I like snakes I like geckos I like crocodiles	Roughly 30
"Tell me about yourself."	My elbow bends both ways I can't keep up with rap lyrics I saw a monkey on the subway I've never tried orange juice	Roughly 1,000,000,000

more information

How do we quantify this intuition? To begin, we enumerate all of the possible messages using binary numbers. In effect, we're translating the messages into a code of 0s and 1s. If there aren't very many messages, then you can cover them all with short codes.

Thus, the fewer digits required, the less information conveyed.

WHAT YOU WERE ASKED	SAMPLE REPLIES	SAMPLE CODES	DIGITS REQUIRED
What reptiles do you like?	I like tortoises I like skinks I like Jem'Hadar I like alligators	00000 00001 00010 00011	5

In contrast, if you're choosing from a long list of possible messages, that will require long codes. Thus, the more digits required, the more information conveyed.

WHAT YOU WERE ASKED	SAMPLE REPLIES	SAMPLE CODES	DIGITS REQUIRED
Tell me about yourself.	I once attacked a windmill	00000000000000000000	
	I have a squire and he's my best friend	00000000000000000001	20
	I have an adjective named after me	00000000000000000010	(or more)
	I dream the impossible dream	00000000000000000011	

Hence, Claude's fundamental unit of information: the binary digit, or "bit."

To see bits in action, let's watch how they come trickling in when we play a classic game of hidden information: hangman.[1] I'll draw from the 2019 Collins *Scrabble Dictionary*, a list of almost 280,000 words. Enumerating them all would require 18-digit binary codes (plus a handful of 19-digit codes). Thus, in Claude's terms, figuring out the word means acquiring roughly 18.09 bits of information. For comparison, that's about a millionth as much as a digital photo.[2]

Now, let's say our opponent writes seven blank spots. That narrows our search down to seven-letter words, of which there are about 34,000. Since we could enumerate these with 15-digit binary codes (plus a handful of 16-digit codes), we have 15.2 bits of information remaining to gather.

That means the information we just gained—that the word is seven letters long—is worth a little under 3 bits.

2.9 out of 18.1 bits

34,342 words left

Now, let's guess our first letter. I like to start with *E*. And hey, good news!

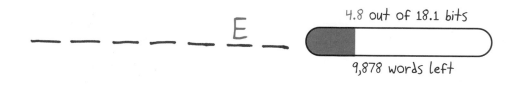

4.8 out of 18.1 bits

E

9,878 words left

---

1   For those who don't know, it's a timeless (and needlessly morbid) game. One player selects a secret word, then writes down a blank space for each letter. The other players guess letters, one at a time, hoping to complete the whole secret word before suffering eight wrong guesses.

2   You could in theory play a version of hangman where you guess a photo's RGB values by pixel, though it might take a few years.

That narrowed the field down to just under 10,000 words, gaining us another 1.8 bits of information. We've eliminated more than 95% of possibilities. Still, as measured in bits, we're just getting started.

Looks like we need another vowel. How about *A*? Alas, bad news.

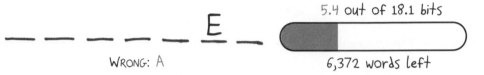

Although we're one step closer to defeat, we at least gained information, eliminating 3,500 words from the list. That's 0.6 bits.

Now, looking at that last blank, I'm wondering: Is it a *D*? Survey says: Nope.

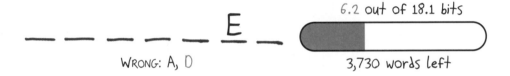

Even so, progress. We've narrowed the field by another 2,600 words, worth 0.8 bits.[3] For the next guess, I'm eyeing that last letter again. How about *S*? Ooh, look at that!

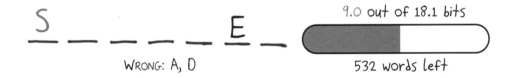

Not what I expected, but a good outcome. That gained us 2.8 bits of information, our best turn yet. Now, we still need another vowel, so how about *I*?

Yes, *I* for the win!

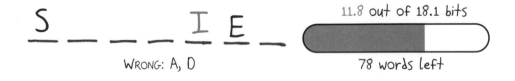

---

3   You may notice a weird contrast in our last two guesses: Whereas guessing *A* eliminated more words, guessing *D* gave us more bits of information. Why? Well, what matters isn't the *number* of possibilities crossed out, but the *fraction*. Guessing *D* eliminated fewer words in absolute terms, but a larger fraction of the words remaining.

That's another 2.8 bits of information. We're down to a few dozen possible words.

It looks like we still need another vowel. How about *O*? Oh. No.

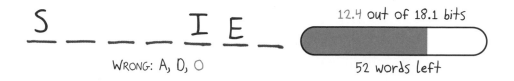

S _ _ _ _ I E _

WRONG: A, D, O

12.4 out of 18.1 bits

52 words left

That's a third wrong guess, and just 0.6 bits to show for it. Let's try another vowel. *U*? Success!

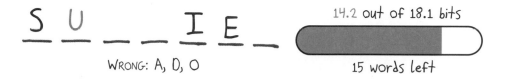

S U _ _ _ I E _

WRONG: A, D, O

14.2 out of 18.1 bits

15 words left

That was worth 1.8 bits, narrowing the field to just 15 possible words. Next, we can turn our attention back to the final letter. Is it *N*? Nuh-uh.

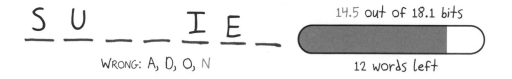

S U _ _ I E _

WRONG: A, D, O, N

14.5 out of 18.1 bits

12 words left

That eliminated just three words, gaining us a meager 0.3 bits of information. Let's try again. How about *R*? Ooh, bonus!

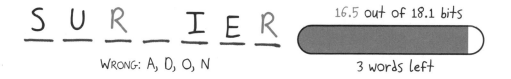

S U R _ I E R

WRONG: A, D, O, N

16.5 out of 18.1 bits

3 words left

That brought us from a dozen possibilities down to just three, delivering precisely 2 bits of information. Next: How about *P*? Oops. Bad guess.

S U R _ I E R

WRONG: A, D, O, N, P

16.5 out of 18.1 bits

3 words left

Since "surpier" isn't a word, that didn't tell us anything. We gained 0 bits of information. Let's be more careful this time: How about *F*? Still no success.

Still, we gained 0.6 bits by eliminating one of the three possibilities, "surfier." We're down to two. A single bit of information eludes us. Shall we try *L*? Ah, there we have it! We win.

Okay, enough of word games. Onward to the math games.

The games in this section are far-flung cousins of hangman. In each, victory requires ferreting out the right information. It may be a secret number (in Bullseyes and Close Calls), an auction item's true value (in Caveat Emptor), a tangled map of regions (in LAP), the identity of mysterious "cards" (in Quantum Go Fish), or the very rules by which you are playing (in Saesara). In each game, one person seeks to divulge as little information as possible, while the other seeks to snatch as many bits as they can.

If suspense is the state of craving knowledge, then information games give us suspense in its purest form. They are mystery novels played out in real time.

What would my onetime neighbor Claude think of his grand theory being repurposed for such frivolities? I'm sure he'd love it. "The history of science has shown that valuable consequences often proliferate from simple curiosity," he once said. Claude should know. When he worked at Bell Labs, he spent whole days in the common area playing board games. His boss said that he had earned "the right to be nonproductive."

That said, you may want to save these games for when you're off the clock, in case your boss isn't as cool as Claude's was.

BULLSEYES AND CLOSE CALLS

A CLASSIC CODE-BREAKING GAME

Under the name Mastermind, this became one of the biggest board games of the 1970s, selling as many units as *The Godfather* sold tickets (about 50 million, if you're keeping score). But the game didn't begin with those colorful plastic pegs. For a century beforehand, it was played using pen and paper, under the earthy name Bulls and Cows. Now, as an ardent bovine feminist, I reject the idea that bulls are better than cows, so I've renamed the former as Bullseyes and the latter as Close Calls. But feel free to use whatever words you wish. This code game, under any code name, remains a stone-code classic.

HOW TO PLAY

What do you need? Two players, pens, and paper.

What's the goal? Guess your opponent's secret number before they guess yours.

What are the rules?

1. Each player writes down a **secret four-digit number**. All of the digits must be different.

2. Take turns guessing four-digit numbers. (Again, no repeating digits in a number.) Your opponent will report how many of your digits were **bullseyes (right digit, right position)**, and how many were **close calls (right digit, wrong position)**. However, **you will not be told which digits were which**.

3. The winner is whoever **scores four bullseyes in the fewest guesses**.

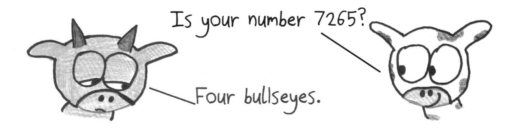

TASTING NOTES

To experiment with strategy, I wrote a simple computer program. First, it began with a list of all possible numbers and made a random guess. Then it crossed out all numbers inconsistent with the feedback received. From those remaining, it made another random guess. This process repeated until the secret was uncovered.

The program did all right, usually requiring five or six guesses. But once in every few thousand rounds, it stumbled its way to an unimpressive nine guesses. Here's one such fiasco:

GUESS	FEEDBACK	NUMBERS LEFT
5873	🐷🐷	1155
3951	🐄🐷	189
2938	🐷🐷	45
3712	🐄🐷	20
8791	🐄🐷	4
8152	(nothing)	3
3097	🐄🐄🐄	2
3497	🐄🐄🐄	1
3697	🐄🐄🐄🐄	solved!

The first five guesses went well, narrowing the field to just four possibilities: 8152, 3097, 3497, and 3697. Then, steps from the finish line, the computer began tripping all over itself, burning four turns to choose among four options.

That's not just bad luck. It's bad strategy, too. If the program had chosen a wiser number for its sixth turn, then it could have guaranteed a solution on its seventh.

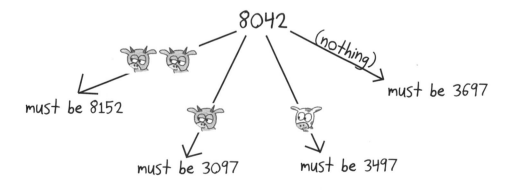

Why didn't the program think of this? Blame its foolish programmer. I had forbidden it from guessing any number it had already eliminated, in effect treating each guess as a chance to win. But a guess is something else, too: a chance to gather precious information.

"To a mathematician," writes teacher Paul Lockhart, "a problem is a *probe*—a test of mathematical reality to see how it behaves. It is our way of 'poking it with a stick' and seeing what happens." To play Bullseyes and Close Calls optimally, you should poke the secret number in the way that gathers the most information—even if it means guessing a number that you know is wrong.

WHERE IT COMES FROM

Like many classic games, its origins are lost to history. All we know is that early 20th-century Britons called the game Bulls and Cows.[4] In the late 1960s and early 1970s, computer versions popped up at Cambridge and MIT. A few years later, Israeli telecommunications expert Mordecai Meirowitz gave it global prominence under the name Mastermind.

WHY IT MATTERS

Because life is a hunt for information, and humans are lazy hunters.

You know this already. Either you're human yourself, or you're conversant enough in human culture to enjoy human books like this one. Either way, you've seen *Homo sapiens* spend hours gorging on information, then somehow emerge from the feast without an ounce of nourishment.

Take a wretched and typical specimen: me. I subscribe to 77 podcasts, follow 600 people on Twitter, and long ago maxed out the number of open tabs in my Wikipedia phone app.[5] Given all this information, how informed am I? The other day my young daughter picked up a pinecone. "That's a pinecone," I volunteered. "It comes from a pine tree. It's . . . some kind of big seed, I think?"

This wasn't a tough pitch to hit. My daughter had not picked up a quasar, a Tom Stoppard play, or the hard problem of consciousness. The truth about pinecones is definitely out there. I just didn't know it. Nine words in, I had exhausted my knowledge.

As a rule, humans don't seek information in the right places. In a classic psychology study, subjects were shown four cards, each with a letter on one side and a number on the other. Then they learned a rule: **A card with a vowel must also have an even number.**

---

4   According to Andrea Angiolino, Italians called it Little Numbers or Strike and Ball.
5   They cap you at 100. My last few: Allie Brosh; list of World Series champions; *The Diary of the Rose*; Ian Haney López; "Big Yellow Taxi"; John Rocker; Santorini (game); Anaïs Mitchell . . . It's a great snapshot of what I've been thinking about recently, in a terrifying, panopticon kind of way.

The question: Which cards do you need to flip over to see if the rule has been violated?

Before reading further, think it over. What would you flip? If you prefer to copy off of other people's homework, here are the most common answers from a typical iteration of the study, conducted in 1971:

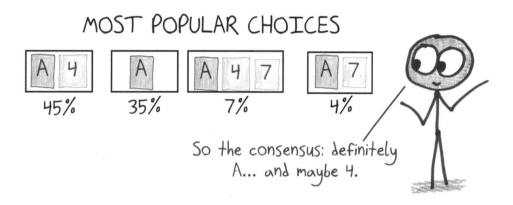

It's pretty clear that we need to flip *A*. But after that, controversy begins. Most people want to flip 4, presumably in order to check for a vowel. But suppose you find *J* or *W* or *P*—who cares? No violation there. The rule said that a vowel must have an even number; it didn't forbid a consonant from having one, too.

Meanwhile, most folks decline to flip 7. It's not even, so the rule is irrelevant, right? Wrong. If you flip 7, and find an *E* or *U*, then the rule has been broken.

The study highlights a pattern called *confirmation bias*. Instead of seeking examples that could challenge our theories, we seek examples to confirm them. Confirmation bias is often blamed on emotion. If I believe Demublicans to be paragons of civic virtue and Republocrats to be monstrous hypocrites, then confirmatory examples will make me feel righteous and superior, while counterexamples will make me feel anxious and persecuted. No surprise which one I seek out.

Still, though emotion plays a role, confirmation bias runs deeper. In the four-card study, folks have no emotional stake in an abstract rule about vowels. There's no benefit to being wrong. Still, 96% fail to give the logically correct answer.

By habit, we look for information in the wrong places, like a space program sending probes to the wrong planets.

Most wrong beliefs cost you nothing. Flat earthers can still buy plane tickets; doubters of the moon landing can still stargaze; you can dislike Outkast and still live an arguably happy life. By contrast, Bullseyes and Close Calls will call you on your crap. Ask a worthless question, get a worthless answer. Instead of letting information wash over us, we must seek it out, probing the world with a sense of purpose—and maybe, when the game is over, looking up what pinecones actually are.[6]

VARIATIONS AND RELATED GAMES

REPEATS ALLOWED: Allow repeat digits, both in the secret number and in guesses. For example, if the secret number is **1112**, and you guess **1221**, then your feedback is **one bullseye** (for the first digit being 1) and **two close calls** (for the final two digits being 2 and 1, but in the other order).[7]

SELF-INCRIMINATION:[8] Every guess is applied to both numbers. That is, when you make a guess, not only does your opponent give feedback, but you give feedback, too, as if your opponent had just made the same guess. For example, if you guess 3456, and your own number is 1234, then you must say

---

6   They're seed containers, offering a layer of protection from the abusive elements.
7   This variant is perhaps more common and natural than the one laid out in the chapter. The only trouble is that it's weirdly hard to articulate what a close call means. To be clear, each digit can only be counted once—for example, if the secret number is 1112, and you guess 2223, that's just one close call, not three.
8   This idea and the next two come from R. Wayne Schmittberger's *New Rules for Classic Games*.

"two close calls." This requires extra strategizing, as you don't want to make a guess that reveals too much about your own number.

SPOT THE LIE: Once during the game, each player may give false feedback—for example, saying "one bullseye, one close call" when in reality there were no bullseyes and three close calls.[9] Try to deploy your lie at the most confounding moment.

TIGHT LIPS: Each guess receives minimal feedback, just "yes, there is at least one bullseye" or "no, there are no bullseyes." Makes for a slower, trickier game.

JOTTO: A word game in the same style. Instead of a four-digit number, each player picks a four-letter word (with no repeated letters). Play proceeds as in Bullseyes and Close Calls, with one extra twist: You may only guess actual words (such as *crab*) not random combinations of letters (such as *racb*). For an extra challenge, use five-letter words instead of four-letter ones.

---

9   The exception: If someone hits four bullseyes, you can't give false feedback, because they just won.

CAVEAT EMPTOR

AN AUCTION GAME

I'm afraid I can't teach you how to win at Caveat Emptor. But I can easily tell you the best way to lose: Just win every auction.

I mean it. Play a few rounds, and you'll find that overbidding is all too common. It's a game marked by Pyrrhic victories, with winners forced to take home prizes for more than they're worth. This phenomenon—losing your shirt on a winning bid—is so pervasive that auction economists have dubbed it "the winner's curse."

Lucky for you, Caveat Emptor offers a lot more information than the typical auction. Will that be enough for you to escape the curse?

HOW TO PLAY

What do you need? Two to eight players; it's best with four to six. Spend a few minutes gathering **five random household objects** to auction off.

Each person also needs **six cards, numbered 1 to 6**. (Scraps of paper will work.) These cards are not used in bidding; instead, they're used to determine each item's secret "true value."

On another sheet, set up a table on which to **track each player's score and the cards that they've used.**

STAR	VILLAIN	ALIEN	PERSON
1	1	1	1
2	2	2	2
3	3	3	3
4	4	4	4
5	5	5	5
6	6	6	6

space for keeping score

for crossing off
numbers used

What's the goal? Win household items at auction, but don't pay more than they're worth.

What are the rules?

1. Each round, one of the players—the auctioneer—picks an item and gives a little speech about how delightful and valuable it is.

Behold! This cuddly carnivore is named in honor of Theodore Roosevelt, who was, of all US presidents, the cuddliest and most carnivorous!

2. Now it's time to determine the true value of the item. To do this, every player (including the auctioneer) **secretly chooses a value from 1 to 6. The sum of these values—which no one yet knows—is the true value of the item on auction.**

teddy bear's true value is the
sum of these secret numbers

3. Next, it's time for bidding. Each player hopes to buy the item for less than its true value. Bidding begins with the player to the left of the auctioneer, who **states a price they'd be willing to pay for the item**.

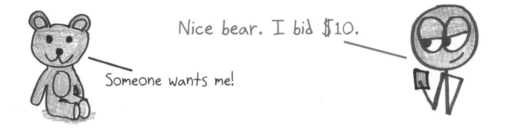

Nice bear. I bid $10.

Someone wants me!

4. Bidding continues to the left. On your turn, you must either **raise the bid** or drop out of the auction. **When you drop out, reveal the value you selected.** Thus, as players drop out, the remaining players gain information about the item's true value.

OPTION 1: RAISE   OPTION 2: DROP OUT

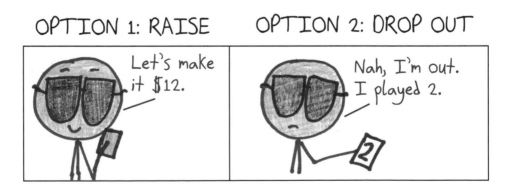

Let's make it $12.

Nah, I'm out. I played 2.

5. If you are the **last player remaining**, then you **win the auction** at the price of your last bid. Reveal your own number. The item's true value is now known to all.

6. **Subtract the price paid from the item's true value, and score this many points**, which may be negative, for the "winner" of the auction. Also, **whatever value you selected, you cannot use again**. Discard those scraps of paper and cross out the corresponding numbers on the scoring table. (You can also accomplish the same result by keeping discarded scraps faceup and visible to everyone.)

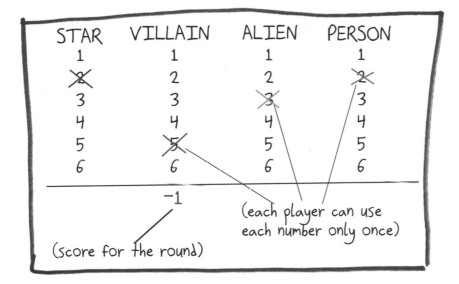

7. **Play five rounds, changing the auctioneer each time.**[10] It's okay if not everyone gets the same number of chances as auctioneer. In the end, the **highest total score wins**.

TASTING NOTES

My favorite part of the game is the speeches. I have been roused to the admiration of a broken pencil and brought to the brink of tears by a peanut-buttered cracker. It seems that everyone becomes a poet when asked to eulogize a coupon for 20% off at Bed Bath & Beyond.

---

10 You can also play a different number of rounds. The trick is that you want to have two cards remaining for the final round, so to play n rounds, start with cards numbered 1 through $n + 1$.

Then, when the speechifying ends, the strategizing begins. Two basic approaches jump out:

1. Select a low value, then act like you selected a high one, so your opponents overbid.

2. Select a high value, then act like you selected a low one, hoping to win the auction yourself.

Still, as each round unfolds and new information trickles in, you'll need to make adjustments on the fly. For example, say a round begins like this:

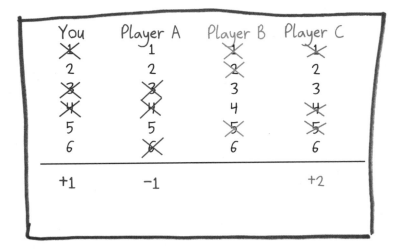

Right away, we can calculate the next item will be worth **at least 8** (that's 2 + 1 + 3 + 2) and **at most 23** (that's 6 + 5 + 6 + 6). After picking your own card—let's say 5—you can update this range. The item must be worth **at least 11** (that's 5 + 1 + 3 + 2) and **at most 22** (that's 5 + 5 + 6 + 6).

Every player is making a similar calculation. So when A opens the bidding at 12, you may suspect that they've played a 5 (which allows for a minimum value of 12) rather than, say, a 1 (which would allow values as low as 8). Or perhaps A is bluffing. Hard to say.

Anyway, let's say B bids 13, and then C drops out, revealing a 2.

THE ROUND SO FAR

I'll let you choose your own adventure. What do you do on your next turn?

YOUR CHOICES

No peeking ahead!

I mean it.

Make your decision first.

Okay, ready?

IF YOU CHOSE OPTION #1 (RAISE): With debonaire confidence, you flash a James Bond smile and raise the bid to 14. This feels smart. This feels right. You're sure that Player A played a high card, so the value must be—

Oh no. Player A just dropped out, revealing a 1. At that, Player B drops out, revealing a 4. That leaves you as the winner of the auction. You just paid $14 for an item worth $12. Not your best work, Bond.

IF YOU CHOSE OPTION #2 (DROP OUT): With a furtive glance around the table, you mutter "I'm out" and reveal your 5. You feel small, skittish, somehow ashamed. Relax, pal; it's only a game.

Anyway, you feel better when Player A drops out, revealing a 1. Having now won with a bid of 13, Player B groans and flips over their card, revealing a 4, and thus the item's total value: $12. Their winning bid was $1 over. Good thing you dropped out!

WHERE IT COMES FROM

Neck-deep in my research on abstract strategy games, I was longing for a change of pace. Something loose, multiplayer, and fit for a party. Something with long speeches on the merits of paperclips. That's when this game occurred to me: Caveat Emptor. Buyer beware.

In the game's early iterations, the winner's curse worked too well. Overbidding was so rampant, you did best by never bidding at all. Clearly, the players needed more information. But what? A play-testing session with friends generated the idea of using each number only once per game, thereby narrowing the possible values in later rounds. I then added the twist of each player revealing their chosen value upon dropping out. Pumping in this extra information brings the winner's curse down to manageable levels.

WHY IT MATTERS

Because everything has a price, and auction winners often overshoot it.

We live in a world on auction. Photographs have been auctioned for $5 million, watches for $25 million, cars for $50 million, and (thanks to the advent of non-fungible tokens) jpegs for $69 million. Google auctions off ads on search terms, the US government auctions off bands of the electromagnetic spectrum, and in 2017, a painting of Jesus crossing his fingers fetched $450 million at auction. Before we dub this the worst-ever use of half a billion dollars, remember two things: (1) The human race spent $528 million on tickets to *The Boss Baby*, and (2) it's a notorious truth about auctions that the winner often overpays.

Why does this winner's curse exist? After all, under the right conditions, we're pretty sharp at estimation. Case in point: In the early history of statistics, 787 people at a county fair attempted to guess the weight of an ox. These were not oxen experts. They were not master weight guessers. They were ordinary, fairgoing folks. Yet somehow their average guess (1,207 pounds) came within 1% of the truth (1,198 pounds). Impressive stuff.

Did you catch the key word, though? *Average.* Individual guesses landed all over the map, some wildly high, some absurdly low. It took aggregating the data into a single numerical average to reveal the wisdom of the crowd.

WHEN DID COLUMBUS REACH THE AMERICAS?

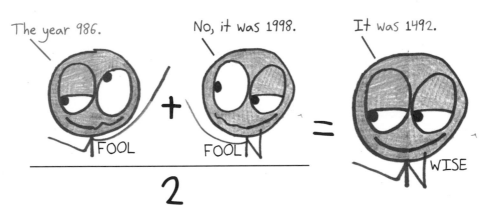

Now, when you bid at an auction—specifically, on an item desired for its exchange value, not for sentimental or personal reasons—you are in effect estimating its value. So is every other bidder. Thus, the true value ought to fall pretty close to the average bid. Here's the thing: Average bids don't win. Items go to the *highest* bidder, at a price of $1 more than whatever the *second-highest* bidder was willing to pay. The second-highest bidder probably overbid, just as the second-highest guesser probably overestimated the ox's weight.

To be sure, not all winners are cursed. In many cases, your bid isn't an estimate of an unknown value but a declaration of the item's personal value to you. In that light, the winner is simply the one who values the item most highly. No curse there.

But other occasions come much closer to Caveat Emptor: The item has a single "true" value, which no one knows precisely, and everyone is trying to estimate.

In such cases, beware victory. It's sometimes worse than defeat.

VARIATIONS AND RELATED GAMES

CAVEAT EMPTOR WITH REAL AUCTIONS: In Caveat Emptor, the auctions are make-believe, with items returned to their original locations after the game. (Please don't try to walk off with someone's childhood teddy bear.) However, Joe Kisenwether suggests a twist: Auction off *actual items*, such as a candy bar, a back rub, or the ability to choose the next game played. The winner really gets to keep it. But I suggest a second twist: You only get the item *if you bid its value or less*. If the winner overbids, no one gets the item.

LIAR'S DICE (AKA DUDO): For this bluffing game from South America, each player needs **five dice and a cup under which to conceal them**. To begin a round, roll your dice, keeping them under the cup. Have a peek, but don't let anyone else see what you've rolled.

Then, the **bidding begins**—for example, five 3s (which means that, across all of the dice, there are at least five 3s). Bidding continues to the left, and each player must raise the bid by **increasing the number of dice** (six 2s), **increasing the value** (five 6s), **or both** (six 4s).

Alternatively, you may **challenge the previous bid** by saying "Dudo" (Spanish for "I doubt it"). At this point, **players all lift their cups to reveal their dice**. If the bid was true, then the challenger loses a die. If the bid was false, then the bidder loses a die. Play then continues, with the loser of each round opening the bidding in the next round—unless they have run out of dice, in which case they are eliminated. **The last player with dice remaining is the winner.**

The game creates a positive feedback loop: The more dice you lose, the less information you control, and the harder it is to make accurate bids (or convincing bluffs). In contrast, the more you win, the greater your control, and the easier it is to win again.

LIAR'S POKER: Similar to Liar's Dice. To play, everyone gets out a **dollar bill** and peeks at its **serial number**, showing no one else.

Then the bidding begins—for example, five 6s (that is, if you count the digits from all of the serial numbers, there are at least five 6s). Bidding continues to the left. Each player must raise the bid by increasing the digit (five 8s), increasing the number of instances (six 3s), or both (seven 9s).

Alternatively, you may challenge the previous bid. A challenge does not force an immediate reveal; instead, the next player may join in the challenge, or simply raise the bid. **Play continues until a particular bid has been challenged by every other player.** At that point, the serial numbers are revealed. If the bid was correct, then every challenger gives the bidder a dollar. If the bid was incorrect, then the bidder must give every challenger a dollar.

I learned the game from *Liar's Poker*, Michael Lewis's scathing memoir of his time on Wall Street. As he tells it, CEO John Gutfreund once challenged investor John Meriwether to a game of Liar's Poker for a stake of $1 million. Meriwether counterproposed a stake of $10 million. It was a bluff, but Gutfreund backed down.

Definitely seems like the culture you want pulling the levers of the world economy.

I cannot condone this game! As money, I prefer to be used for more wholesome purposes, such as organized crime.

LAP

A GAME OF LABYRINTHINE AREA PUZZLES

Hey, you know Battleship? Well, forget it. Wipe it from your memory. Delete the file.

Hey, you know Battleship? No? Good, the memory wipe worked. Now you're ready to meet a better game, a harder game, the game that Battleship aspires to be: LAP. Its name comes from the initials of its creator, Lech A. Pijanowski, but it could just as easily stand for "Let's All Play," "Labyrinthine Area Puzzles," or "Like a Pro." As in Battleship,[11] players probe a hidden grid. Yet LAP is deeper, subtler, and ultimately more rewarding. So, let's all play labyrinthine area puzzles like a pro.

HOW TO PLAY

What do you need? Two players, each with a **6-by-6 grid**, **plus extra grids** on which to track information gained from your opponent.

What's the goal? Fully map your opponent's regions before they can map yours.

What are the rules?

---

11 Oh, you've never heard of it? Well, it's like LAP, but not as good.

1. To begin, **secretly divide your grid into four equal-sized regions:** I, II, III, and IV. A region consists of precisely nine connected squares. Diagonal connections don't count. I suggest distinguishing them three ways: numbers, shading patterns, and pencil color (though in theory, just numbers would suffice).

2. Take turns **asking your opponent about a rectangle of squares** (e.g., B3 to C4). The rectangle must be at least 2 by 2, but it may be larger. Your opponent will tell you **which regions those squares belong to** (e.g., I, II, IV, IV), **but not their formation** (e.g., you don't learn *which* square belongs to region I).

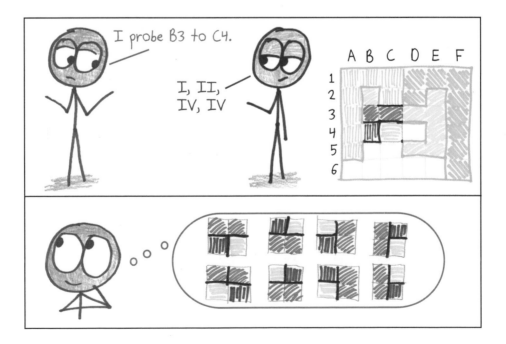

3. When you have solved your opponent's board, **announce that you are making a guess**. Then, hold your guess and your opponent's board side by side. If they are identical, you win. If not, you lose.

WINNER

guess   =   truth

(colors need not match, just regions)

guess   ≠   truth

WINNER

mismatch!

TASTING NOTES

The name of the game is extracting information. Well, not actually; the name of the game is Little Affable Probes. Still, extracting information is the challenge and thrill of the game, if not its title.

I like to begin by probing the corners. For example, if I'm told the upper left corner is three Is and one II, then I know it must be one of three formations:

impossible; region II would not be connected

Next, I might learn that the upper-right corner is all Is. I can then deduce that the Is hug the top wall. If they didn't, then connecting the Is would require them to wrap around region II, which is impossible. Try it: Either region II would have to be smaller than nine squares, or region I would have to be larger.

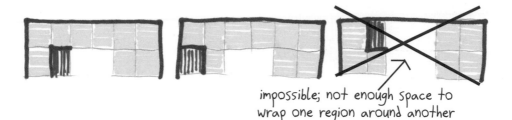

impossible; not enough space to wrap one region around another

LAP offers a classic trade-off. Do I seek the *most* information or the *easiest-to-interpret* information? Myself, I pick the corners first, because the logical deductions are clear. But expert player Bart Wright (who plays on an 8-by-8 board) prefers to begin with the center, knowing that from this information, he can usually deduce the values along the edges. Thus, I play for easy-to-interpret information, while Bart plays for maximal information.

There are also strategic considerations in designing your board. Regions with big solid chunks are easier to figure out, and certain board designs are especially devious:

In Lech's original rules to LAP, which only allowed 2-by-2 probes, the four boards above would be indistinguishable. That's why I've tweaked the rules to allow larger rectangular probes; they let you resolve this ambiguity.

WHERE IT COMES FROM

The game was first published in a Polish newspaper column written by the game's creator, Lech Pijanowski. He later took the leap of sharing it with celebrated game designer Sid Sackson. "It could be a shock, really," Lech wrote to Sid, "getting a letter like this one, from a faraway country and an unknown person." Lech needn't have worried. Sid was so impressed, he translated the game from Polish ("with considerable difficulty") and featured it in his hit 1969 book, *A Gamut of Games*. LAP does the things that Sid said a good game should: "be easy to learn yet have infinite strategic possibilities, give you a chance to make choices, create interaction among players[12] and take a maximum of one and a half hours to play."

It's clear that Sid was fond not only of the game but also its creator. "In another twenty years," he wrote, "we might begin to exhaust the information we have to impart to each other." Alas, Lech died just a few years later. Sid outlived him by 28 years.

WHY IT MATTERS

Because no information exists in isolation.

Tell me if you've had this experience. A person says something false, like "The moon is a hoax." "You mean the moon landing?" you ask. "No," they insist, "the moon itself." Spurred to act, you give a heroic succession of patient arguments until, finally, you persuade them of the truth. Then, months later, you run into the person again . . .

And they're back to spouting the same old nonsense.

What went wrong? Have you time-traveled into the past, so that your decisive argument has yet to take place? Did all that hard-won insight just vanish?

According to psychologist Jean Piaget, we respond to novel information via two basic processes. The smoother one is *assimilation*: adjusting new facts to fit your existing worldview. The rougher one is *accommodation*: adjusting your worldview to make room for challenging new facts.

12 Well . . . you could dock LAP points here. Its style is what gamers sometimes call "simultaneous solitaire." But not all games can be all things to all people, with the possible exception of "the floor is lava."

LAP is a toy model of this process. For example, say I'm told that the upper-left corner contains one I, one II, and two IIIs. Based on that fact alone, 12 formations are possible. However, while my gameboard might be a blank slate, my mind isn't. I have a preexisting worldview; in particular, I know each region must be connected. That rules out any formation with an isolated square in the corner.

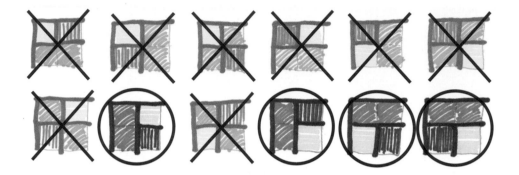

In absorbing the new information, I modify and reinterpret it, winnowing 12 possibilities down to four. That's how assimilation works. Filling a mind with facts is not like filling a vessel with water. For anyone with a worldview—that is, anyone with a pulse—assimilation is an active process.

As for accommodation, it occurs when new information forces me to confront a mistake. Say I've deduced—or believe that I've deduced—the top two rows of the board. Then I probe C3 to D4, and learn that it contains three Is and a III.

This is, according to my worldview, impossible. One of my beliefs must give way.

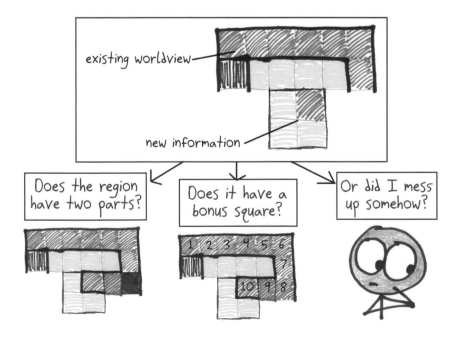

In life, as in LAP, knowledge is not just a pile of logical propositions. It's a network of them, held together by beliefs, experiences, and values. Often, when we fail to learn something, it's because we struggle to reconcile the new information with the old. Assimilation fails, accommodation is scary, and so the incoming data, no matter how factual, is rejected like a foreign pathogen.

Want to persuade your fellow human beings of the truth? It's hard. You've got to speak to their worldview. You've got to help them see how they can revise particular beliefs while keeping intact their values, identity, and sense of self. No easy feat. And to build that trust, it can't hurt to start with a friendly game of LAP.

VARIATIONS AND RELATED GAMES

BEGINNERS' LAP: Divide your 6-by-6 grid into just two regions.

EXPERTS' LAP: Play on an 8-by-8 grid, divided into four regions. This is how the game was originally published. With 64 squares instead of just 36, it takes quite a bit longer.

CLASSIC LAP: You must probe a 2-by-2 box each time; no larger rectangles are allowed. (The unguessable boards shown under "Tasting Notes" are forbidden.)

RAINBOW LOGIC: Math educators Elizabeth Cohen and Rachel Lotan, in their book *Designing Groupwork: Strategies for the Heterogeneous Classroom*, give a simple and elegant variation on LAP. Play on a 4-by-4 board, with four equal-sized regions. But instead of asking about the contents of a 2-by-2 rectangle, **probe the contents of a particular row or column**. For an extra challenge, try five regions on a 5-by-5 board.

QUANTUM GO FISH

A GAME OF MYSTERIOUS FINGERS

As we begin this chapter, conscience compels me to confess an error. I'm sure this book has many, but only one of them was a knowing and deliberate lie. Remember when I said Quantum Tic-Tac-Toe was the trickiest game in the book?

The real champion is this glorious monster. I consider it a cross between a logic puzzle, an improv comedy session, and a collective hallucination, played with the strangest deck of cards you've ever seen. To be candid, I'm still wrapping my head around it. Anyway, for a book that started with a childhood game, I can think of no better culmination than this doozy, whose primary fan base is math PhD students.

HOW TO PLAY

What do you need? Anywhere from three to eight players. Each begins the game by holding up four fingers. These are the "cards" in the deck.

"cards"

What's the goal? There are two ways to win:

1.   Prove that you have four cards in the same suit.

2.   State exactly what suits every player has in their hand.

What are the rules?

1. To begin, **no one knows the suits of their own (or anyone else's) cards**. All is a mystery. We only know that there are **four cards per suit**, and as many suits as there are players.

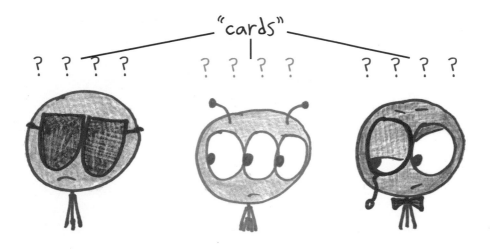

2. On your turn, pick another player, and **ask them if they have any cards from a particular suit**. The first person to reference a new suit makes up a silly name for it. Note that **you may only ask for a suit you already possess**; thus, by asking for Unicorns, you are committing one of your as-yet-unknown cards to being a Unicorn.

3. The asked player can respond in one of two ways:
   - **"No, I don't have any."** Thus, all their cards must belong to other suits.
   - **"Yes, here is one."** In this case, they give **precisely one card** to the asking player. Their other cards remain a mystery (and may or may not belong to the same suit).

4. **Sometimes this decision will be forced.** For example, if you've already committed to having Radishes, and I ask you for Radishes, then you must give me one. **If not forced, the asked player may respond whichever way they wish.**

5. You can win in either of two ways:
    - At the end of your turn, **state exactly what cards each player must have**.
    - At the end of your turn, **prove that you have four cards in the same suit**.

6. However, if a paradox occurs—meaning that the players collectively possess five or more cards all in the same suit, and no one caught the error at the time—then **something has gone wrong, and everybody loses**.

TOO MANY CORNCOBS; EVERYONE LOSES

TASTING NOTES

I myself had no idea how this game worked until I saw a sample round, so here's a step-by-step illustration with three players. They begin with 12 cards: four each of three different suits. No one knows what cards they (or anyone else) has.

Xia goes first and asks: "Yael, do you have any Narwhals?" Yael chooses to reply "No."

Yael goes next and asks: "Zoe, do you have any Scruples?"
Zoe chooses to reply "Yes."

The result leaves Zoe with three cards, and Yael with five. Of these, two must be Scruples: one implied by asking, and one gained from Zoe.

Zoe goes next and asks: "Yael, do you have any Qualms?"

Yael may feel tempted to say "No." But this would lead to a game-destroying paradox. Yael has three cards which are not Narwhals; if they're not Qualms either, then they must be Scruples. That would give Yael five Scruples, which is an impossibility.

Therefore, Yael must say "Yes" and give a Qualm to Zoe.

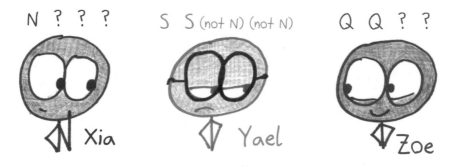

Xia goes next and asks Zoe, "Do you have any Narwhals?"

Clever move. If Zoe says "No," then neither Yael nor Zoe would have any Narwhals. Thus, Xia would have them all, and could claim victory.[13] Instead, Zoe says "Yes," and gives a Narwhal to Xia, who now possesses at least two Narwhals.

---

13 Some players forbid starting with four cards all of the same suit; under that rule, Zoe would be required to say "Yes" here.

Yael, with the next turn, asks, "Xia, do you have any Qualms?" This means that one of Yael's remaining cards must be the third Qualm.

Xia chooses to reply "Yes" and gives the final Qualm to Yael. All the Qualms are now spoken for. Moreover, since Yael's final card can't be a Narwhal, and can no longer be a Qualm, it must be a Scruple.

The next turn falls to Zoe who asks, "Xia, do you have any Scruples?"

This means Zoe's final card is a Scruple. Indeed, it's the last Scruple, which means that Xia can't possibly have one. Why did Zoe even bother to ask?

Because, with all the Scruples and Qualms accounted for, Zoe knows that Xia's remaining cards are Narwhals. By declaring and explaining this knowledge, Zoe wins the game.[14]

Easy like Sunday morning, right?

Some mathematicians I know like to forbid pencil and paper, forcing you to keep track of the game entirely in your head. "Though that's fun," says Anton Geraschenko, "I came up with a **physical deck** for playing the game which automatically does a lot of the bookkeeping for you, freeing your brain cycles up to strategize."

I heartily recommend Anton's system. Here's what you need:

1. **Four paper clips** per player (representing your cards).
2. Assuming you have _n_ players, each player needs **faceup pieces of paper numbered 1 to _n_** (representing the possible suits that you might possess).

---

14 Xia, despite ending the game with all four Narwhals, began with only three, later gaining one from Zoe.

As you play, keep track of the evolving game state via these steps:

1. If you determine that you have none of a suit (because you've answered "No" or because others have them all), **turn the corresponding piece of paper facedown**.
2. Your **unattached clips may belong to any of the faceup suits**.
3. If you determine a card's suit, **attach that paper clip to the corresponding piece of paper**. If you have multiple of that suit, attach multiple clips.

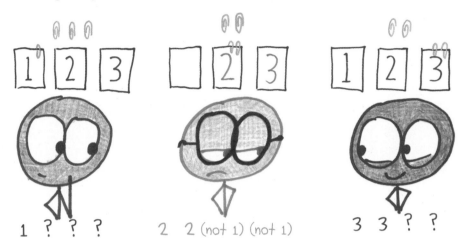

WHERE IT COMES FROM

The game has circulated for years among mathematicians. I've adapted my description from Anton Geraschenko, who encountered the game in the math PhD program at UC Berkeley, where our mutual pal David Penneys was its biggest cheerleader.

Folks there often played it as a drinking game, which I find (1) kind of insane, and (2) kind of inspiring. Whereas most drinking games create runaway positive feedback cycles (losing makes you drink, which makes you drunk, which makes you lose, which makes you drink . . .), this one is played with a healthier *negative* feedback loop, in which someone who makes an error is forbidden from drinking for the next round.

Anyway, Anton learned it from Scott Morrison, who credits Dylan Thurston, who doesn't know who invented the game but shared with me his earliest record of it: a 2002 email, signed by himself and Chung-chieh Shan. That version (called Quantum Fingers) differed in a few ways: (1) When you complete a suit, you lower those four fingers, but the game does not end. (2) To win, you must lower all of your fingers. (3) No one can start with four cards of the same suit.

WHY IT MATTERS

Because mathematical games unlock powers we didn't know we had.

When I was nine years old, I received a life-changing gift, one that awoke in me a kind of magic: Rush Hour. Not the action movie (though the chemistry between Jackie Chan and Chris Tucker *is* magical); I mean the set of colorful plastic cars and trucks. It came with a 6-by-6 grid and a deck of puzzle cards, each depicting a traffic jam in which to arrange the vehicles. The goal is to slide the pieces around until the special red car can escape through an opening on the edge.

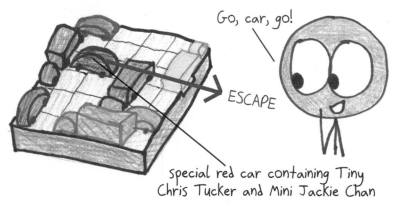

Go, car, go!

ESCAPE

special red car containing Tiny Chris Tucker and Mini Jackie Chan

When I tried to solve the puzzles by imposing top-down logic, I always failed. My brain felt like the equivalent of a 1980s personal computer, too antiquated to run the necessary software. But when I stopped thinking and just *moved*, it all fell into place. My fingers performed dances that my conscious mind could never have choreographed. I solved the puzzles without knowing why or how, answers blooming like time-lapse videos of flowers.

It was my first taste of a startling truth: The mind has powers beyond its own awareness. Some of our intellectual wealth lies hidden in off-the-books accounts.

Quantum Go Fish is the hardest finger game that I know. On first encounter, I doubted that my lumpy apish brain could juggle all the necessary information. Then again, wouldn't I have said the same of Rush Hour? Isn't there similar magic in a poker player reasoning about unknown cards, a chess master intuiting threats and opportunities, a sudoku wizard deducing the next number in a fraction of a second?

Haven't games always driven us to greatness?

This is our human inheritance. We are the monkeys that climbed out of the trees to play hide-and-seek. We are the Peter Pans of the primate order, the chimpanzees who never grew up. We play and we play and we play, and only when our hearts stop beating do we step aside to let others carry on the game.

I've forgotten most of high school chemistry,[15] but I remember this: In metals, all of the atoms pool their electrons together. The electrons flow around

---

15 I'm sorry, Ms. Jackman; I am for reeeeeal.

the whole substance, a shared reservoir of electric charge. That's the clearest image I have for how games work. There's a kind of pooling of energy, a current flowing from hand to eye, from opponent to opponent. The shared anticipation of each other's moves creates something unspoken, almost telepathic.

Quantum Go Fish exhibits this crude primate telepathy better than any game I know. It is a shared fiction, written one question and one finger at a time. Like the flow of electrons around a metal, it follows logical rules, yet it proceeds in bursts and sparks.

This is why math teachers like me prize games so highly. It's not that they're fun (though they are) or that they illustrate key concepts (though some do) or that they fill awkward lessons before vacations (though they have saved my skin a few times). It's that too often, classroom math asks us to reason alone, and games compel us to reason together. That's how we become our best. Our most electrified. Our most human.

And besides, who doesn't want to call their fingers "stellar anomalies" and "bonbons"?

VARIATIONS AND RELATED GAMES

LOSE A TURN: I prefer to play that paradoxical answers are impossible, like trying to move a rook diagonally. If you try it, the other players will stop you, and you just move again. But if you prefer, you can play a more cutthroat version: Anyone giving a paradoxical answer loses the game, and as a penalty, must sit out the next round.

PLAYING ON: If you achieve four cards in a suit, lower those four fingers. The only way to win the game is to run out of fingers.

BLIND KWARTET: I learned of this game from Vincent van der Noort. In the Netherlands, the game Kwartet works much like Go Fish, except that you must ask for specific cards. For example, instead of asking, "Do you have any Fruits?" you would ask, "From the Fruits, do you have the Banana?" (The other fruits might be Apple, Mango, and Kiwi.) Of course, you may only ask for the Banana if you yourself have a Fruit.

Blind Kwartet extends this principle. For example, I might begin by asking, "From the '90s Bands, do you have Chumbawamba?" If you answer "No," then you might still have other '90s Bands, such as Eve 6 or Third Eye Blind. Labeling each individual card gives extra chances for silly fun. But watch out for contradictions: If you're holding a '90s Band and four cards in that suit have been named, then yours must be among them.

SAESARA

A GAME OF INDUCTION

Time for a quick vocabulary lesson. Here are two key terms in philosophy.

DEDUCTIVE REASONING builds chains of logic, moving from general laws to specific conclusions.

ALL MEN DIE

Humans are mortal... and Socrates is human...

Socrates! You gonna die, man!

Rude.

INDUCTIVE REASONING identifies patterns, moving from specific evidence to general rules.

Socrates died... Plato died... Aristotle died...

Do ALL philosophers die?

Now, comprehension check: Which kind of reasoning do most games promote?

The answer is deductive. We know the rules of chess from the outset, and the strategic challenge lies in applying them to new situations. By contrast, inductive games—where you seek to ferret out unknown rules—well, those are rare. Not only rare, but special. Not only special, but a riveting model of scientific inquiry. And not only a riveting model of scientific inquiry, but pretty darn fun, too.

HOW TO PLAY SAESARA

What do you need? Three to five players, each with a pen or pencil. Also, for each round, an 8-by-8 grid with amply sized squares.

What's the goal? Figure out the secret rule for placing numbers.

What are the rules?

1. To begin the round, one player—**the patternmaker—comes up with a secret rule for writing numbers on the grid.** They may also choose to **start the game with a 0**. (The 0 is not required, unless your rule needs a "previous" number to play off of.)

2. Other players then take turns attempting to place marks by pointing a pencil at a square and asking the patternmaker, **"May I place a number here?"** If the patternmaker says yes, **write the next number**. If the patternmaker says no, don't write anything.

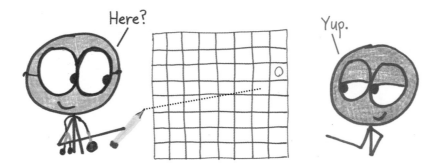

3. On any turn, after trying to place a number, you may also **try to guess the rule**. If you're wrong, **the patternmaker must demonstrate**, by showing either (a) an allowed move that your rule would have forbidden or (b) a forbidden move that your rule would have allowed. No other hints or feedback are permitted.[16]

---

16 If necessary, the patternmaker may cite counterexamples from the past ("Your rule would have forbidden a number here, but look, we already placed one") or the future ("Your rule works for the next number, but it doesn't apply to the number after; see, here's a counterexample"). But if a rule correctly describes all past turns, and correctly predicts all future ones, then it's the correct rule, even if it's not phrased as the patternmaker imagined.

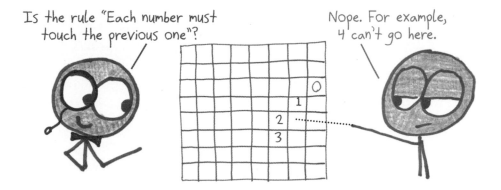

4. If **you guess the rule correctly**, then the round ends. Divide the highest number on the grid by 2 (rounding down if necessary) and give this number of points to **both the correct guesser and to the patternmaker**.

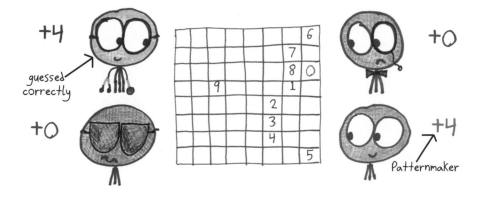

5. However, the round can end in other ways: (1) **if the board reaches the number 20 without the rule being guessed**, (2) **if it becomes impossible to place any more numbers**, or (3) **if one guesser says "Should we give up?" and the others all agree**. In any such case, the round is considered a stalemate, and **everyone scores zero**.

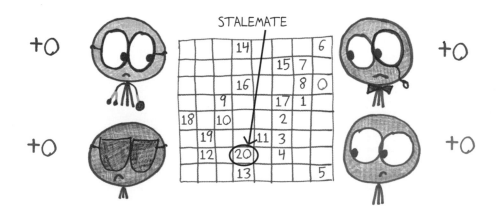

6. Play until everyone has had **one turn each as patternmaker** (or, if you prefer, two turns each). Highest total score wins.

NOTES ON RULE DESIGN

I can't emphasize this enough: **Make your rule guessable!** Your turn as patternmaker is a great chance to score points, and that's wasted on a stalemate. Rules should not be too complex, too weird, too restrictive (lest you run out of legal moves), or too permissive (lest you reach 20 marks without a successful guess). Remember that **rules are always, always, *always* harder to guess than you'd expect**.

All that said, your rule can include any combination of these factors:[17]

·   **Board geography.** "If the grid were colored like a checkerboard, then you may only place numbers on the black squares."
·   **The number itself.** "Odd numbers on the top half of the board; even numbers on the bottom half."
·   **The previous number.** "Each number must be in a different row and different column than the previous number."
·   **All previous numbers.** "Each number must be touching exactly one previous number."

TASTING NOTES

In Saesara, players collaborate to gather data, hoping to uncover a rule that governs and explains everything they've seen. It's just like science at its best. Also, only one person can get credit, just like science at its not-best.

How do you stay ahead of your competitors? Well, remember that guessing the rule isn't just a chance to win. It's a chance to gather information. Proposing a rule like "you can't go anywhere" will force the patternmaker to show you a valid move. Meanwhile, guard your hunches carefully: Guessing a rule that's almost but not quite right could hand the round to your opponent. It's safer to test your hypothesis a few times before venturing a public guess.

As for the patternmaker, you are a scientific law that wants to be found, but not too quickly. How do you manage that trick?

Well, when a player guesses an incorrect rule, calibrate your feedback. Early on, to drag the round out, give minimally informative counterexamples. Later, to avoid stalemate, give maximally informative counterexamples, ones that highlight your rule's essential features.

---

17 I recommend against rules based on invisible factors, such as who placed the number, or whether that square has been tried before. But sufficiently expert players may ignore this advice, as they tend to do with all advice.

WHERE IT COMES FROM

Saesara emerges from a distinguished family tree of "guess the rule" inductive games. Its great-grandparent is Robert Abbott's 1956 card game Eleusis, in which you attempt to uncover the dealer's rule for which cards are playable. Abbott's game hatched several offspring, including John Golden's nifty, Abbott-endorsed Eleusis Express; Sid Sackson's Patterns II (featured in the next chapter); Kory Heath's Zendo (the masterwork of the genre); and Saesara's most direct ancestor, Eric Solomon's pencil-and-paper game that he confusingly *also* called Eleusis. I've reworked Solomon's scoring system, switched letters for numbers, and turned it from a game of mark placing into one of rule guessing. That feels like enough to justify a new name. To stick with the theme, I've chosen Saesara, the ancient name of the Greek city of Eleusis.

WHY IT MATTERS

Because it's the essence of scientific thinking.

Over the past few centuries, science has reshaped the world in its image. That's not because scientists think better thoughts than the rest of us.[18] Nor do they necessarily think *more* thoughts.[19] What makes scientists special isn't the thinking itself, but what happens next.

They try to prove their own thoughts wrong.

Here's the process. First, you formulate an idea about the world. Second, you develop from that idea a set of concrete predictions. Third, you test those predictions via experiment. And finally, you begin the cycle anew, exploring how your original idea withstood the empirical challenge.

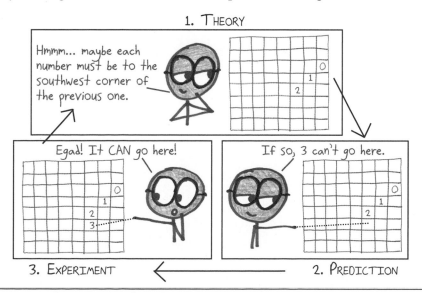

---

18 Charles Darwin, in a letter to a friend: "I am very poorly today & very stupid & I hate everybody & everything."
19 Albert Einstein, asked if he kept a notebook handy to record his ideas: "Oh, that's not necessary. It's so seldom I have one."

Sound simple? Yes and no. Each step is essential, and subtler than it seems.

First, it's deadly—and all too common—to skip step 1 and begin without an idea. You gather data first, and later you contrive a theory to fit the results. The trouble is that the resulting hypothesis will *feel* true, like the fruit of a complete scientific cycle, even though it hasn't really been tested. After all, you can always find some kind of pattern, even in a shuffled deck of cards.[20] It doesn't mean the pattern will hold up on the next shuffle. By skipping straight to step 3, you actually remained stuck on step 1.

Second, it's tricky to find the right predictions. We tend to seek confirmation: "My theory said X would happen, and it did." But if 17 other theories predicted the same result, then such evidence is worthless. Instead, you've got to drive a wedge between possible explanations, to construct an obstacle course that your theory—and your theory alone—could survive.

20 I just tried it, and found that when two consecutive hearts, spades, or diamonds appear, the second card is always lower. *Annals of Mathematics*, DM me!

Third, gathering data is no afterthought. Saesara makes it easy to gather data; just point to a square. But in real science, it's 90% of the battle. Economists can't randomly assign interest rates to countries, physicists can't replicate the Big Bang in a lab, and psychologists struggle to find subjects who aren't undergrads taking their classes. I'm not saying that experimentation is harder than theory, per se. But it's worth noting that Einstein theorized gravitational waves in 1916, and they weren't detected empirically until 2015.

Okay, maybe I *am* saying that experimentation is harder than theory. To be precise, it's 99 years harder.

Textbooks will tell you that math is a deductive subject. In fact, it's *the* deductive subject, the exemplar and epitome of top-down, rule-governed thinking.

But who listens to textbooks?

"Creativity is the heart and soul of mathematics," wrote R. C. Buck. "To look at mathematics without the creative side of it, is to look at a black-and-white photograph of a Cézanne; outlines may be there, but everything that matters is missing."

Just like scientists, mathematicians spend their days playing with new ideas, testing out hypotheses, and running pencil-and-paper experiments. Saesara, like its fellow inductive games, speaks to this unheralded side of mathematical work. As Martin Gardner wrote of Eleusis, it brings out "precisely those psychological abilities in concept formation that seem to underlie the 'hunches' of creative thinkers."

Saesara is a game for the creative mathematician, the pattern-seeking mathematician, the experimental mathematician. It's a game, if you will, for the inductive mathematician.

VARIATIONS AND RELATED GAMES

SPEED SAESARA: Play on a **6-by-6 board** and lower the **stalemate threshold to 10**. It's a faster game that pressures patternmakers into easier rules.

GRAND SAESARA: Play on a **10-by-10 board** and raise the **stalemate threshold to 30**. A longer, slower game, allowing for more mysterious and complex rules. Beginners beware!

JEWELS IN THE SAND: For two to eight players, this is the simplest (and perhaps the most elegant) inductive game around. One player, the judge, makes up a **secret rule for distinguishing jewels vs. sand**. The judge then provides the other players with the following information:

1.   The **category** of objects to classify (e.g., numbers)
2.   An **example jewel** (e.g., 2,000)
3.   An **example of sand** (e.g., 7)

On your turn, name an object, and **ask either "Is it a jewel?" or "Is it sand?"** If the judge says **"Yes," then you keep asking questions**. If the judge says "No," then your turn ends.

At any point during your turn, **you may attempt to guess the rule** (e.g., "Numbers 100 and up are jewels; numbers below 100 are sand"). If you're wrong, then the judge demonstrates by giving a counterexample (e.g., "12 is a jewel" or "9,999 is sand"), and your turn ends. If you're right, then you win, and **serve as judge for the next round**.

Some suggestions from Andy Juell, suited to pretty much any class in school:

  •   **Chemistry:** Mercury and bromine are jewels; iron and helium are sand.
  •   **English literature:** "Quickly," "yesterday," and "here" are jewels; "myself,'" "bicycle," and "green" are sand.
  •   **History:** Fort Sumter and Pearl Harbor are jewels; Gettysburg and Midway are sand.
  •   **Music:** D and G major are jewels; C and F major are sand.

Andy tells me he has already forgotten what rules he had in mind for these, so you are free to let your imagination run wild.

A DISPATCH OF INFORMATION GAMES

You could argue that every game is an information game. Each turn is a kind of signal, and the board is a kind of low-fidelity phone line for exchanging a specialized form of data. Well, you *could* argue that . . . but why would you? Everyone knows that the real information games are these juicy peaches.

BATTLESHIP

A GAME OF TANGY TERMINOLOGY

Decades before Milton Bradley's plastic boards-and-pegs version, Battleship made its first mark as a pencil-and-paper game. "With the new game a new terminology has hit the country," raved one newspaper. "It is tangy and nautical, picturesque and amusing. 'Your salvo,' 'Hit me amid-ships,' 'Got me in the cruiser,' 'I've got your range,' and many other like phrases lend zest to the game."

Of the many ways to tweak the rules, here's my preferred way to play:

1. Each player draws a **10-by-10 grid**, and **secretly places five "ships"** (of lengths 2, 3, 3, 4, and 5) by shading in consecutive squares within a row or column.

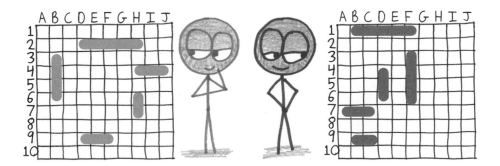

2. Take turns **"firing missiles"** at each other's grids by naming **three squares**. Your opponent will report **how many of these three were hits, but not which ones.** Chart the feedback you receive on **a blank 10-by-10 grid.**

3. When all squares of a ship have been "hit," it sinks. You must **report when each of your ships is sunk**, mentioning its length as you do so.

4. First to **sink all of their opponent's ships** is the winner.

QUANTUM HANGMAN

A GAME OF SIMULTANEOUS WORDS

In classic hangman, players guess one letter at a time, aiming to figure out a secret word before committing eight wrong guesses. In this sneaky variant, suggested by Aviv Newman, you pick two words of the same length (like "skunk" and "apple"). Other players then guess letters, with the following outcomes:

1. **If the letter is in neither word, then the guess counts as wrong.**[21] If the letter is in either or both words, then **all corresponding blanks are filled.**

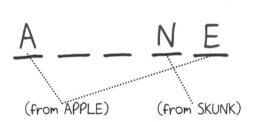

2. At some point, **two conflicting letters will occupy the same blank**. When this occurs, the guessers must "collapse the waveform" by **choosing which letter to keep**.

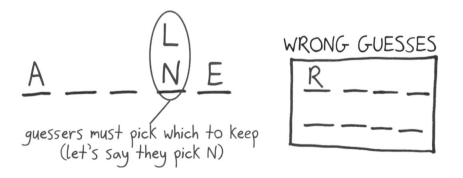

---

21 Usually, these wrong guesses are marked by drawing the limbs of a stick figure being hanged. I prefer my children's games without gruesome illustrations of execution, thank you.

3. One of the two words is thus eliminated, and all of its letters are stricken
    from the board. This may result in **some letters becoming wrong
    guesses retroactively**.

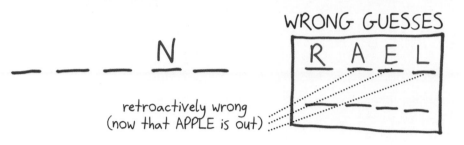

retroactively wrong
(now that APPLE is out)

4. From there, **play proceeds as in normal hangman**. If the guessers
    reach eight wrong guesses, they lose; if they guess the word before that,
    they win.

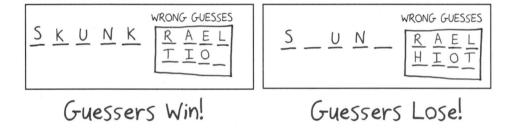

Guessers Win!   Guessers Lose!

For a next-level game, try playing with *three* simultaneous words. When
the first conflict occurs, the player chooses a word to eliminate. Later, another
conflict will occur, at which point the final word will be determined.

BURIED TREASURE

A GAME OF BLUFFING (BUT NOT LYING)

I learned this traditional two-player game from Eric Solomon's *Games with Pencil and Paper*. "It introduces the idea of bluff," he notes, "without requiring the players to lie." Perfect for the child who is honest in deed yet sneaky in spirit.

To begin, write the **letters *A* through *I* on individual scraps of paper**. Without looking, randomly give **four scraps to one player, and four to the other**. Then do the same with **the numbers 1 through 9**. Look at your own scraps, but keep them secret from your opponent.

One number and one letter will remain left over. Like the unclaimed cards in Clue, these unknown scraps of paper **designate the square where the treasure is buried**.

On each turn, take two actions:

1. **Ask your opponent if they hold a particular letter or number.** They must tell the truth. Here's where the bluffing comes in: You may throw your opponent off the scent by asking about a number that you yourself already possess.
2. **Propose a location to dig for the buried treasure.** If your opponent holds either card, then they reply "No treasure there." They need not reveal *which* card they hold.

YOUR TURN, PART 1   YOUR TURN, PART 2

If **your opponent holds neither of the cards from your dig location**, then either (a) you were bluffing and hold one of the cards yourself, or (b) you've found the right spot. If it's the first, say, "Actually, no treasure there." If it's the second, then you win and can claim your prize. (Make sure to check the hidden cards first, to ensure there hasn't been a mistake.)

Eric Solomon notes that the ideal treasure is "something tangible like a toffee apple." It's true: All things, including cars, beaches, and video game consoles, would be better if they were toffee apples.

PATTERNS II

A GAME OF SECRET MOSAICS

Just as the 1990 horror film *Troll 2* bears no connection to 1986's *Troll*, this game from Sid Sackson is not really a sequel. The game stands on its own, a pattern of one. It requires three players, and it is best with four or five.

To begin, the **designer creates a secret pattern by filling a 6-by-6 grid** with any arrangement of four symbols. Other players will attempt to determine this pattern by requesting information, hoping to do so with as few hints as possible.

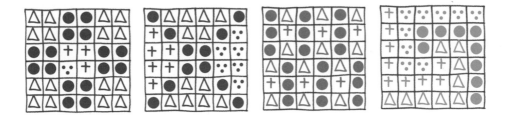

Each guesser begins with a blank 6-by-6 grid. **To receive information, mark as many squares as you wish** with a little line in the lower left corner and pass your sheet to the designer. The designer will fill in the specified squares and secretly pass the sheet back to you. This process repeats as many times as you like, as rapidly as you like; there are no pre-specified "turns."

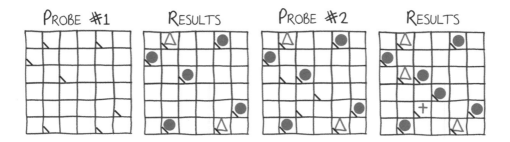

Then, when you believe you've figured out the pattern, **fill in predicted symbols** for as many of the remaining squares as you wish (leaving some blank if you like). After all players have made predictions, the designer scores them, giving **+1 for each correct prediction and –1 for each incorrect prediction**.

Thus, a daring player might venture predictions based on fairly little information, risking a negative score to have a chance at a big one. Meanwhile, a cautious player might prefer to gather lots of information, then make confident guesses on a few final squares.

For their score, the designer receives the **difference between the highest and lowest scores**. Thus, the ideal design achieves a big difference: easy for one guesser, hard for another.

Instead of guessing, **a player may simply give up**, accepting a score of 0 for the round. The designer loses 5 points for the first player who gives up, and 10 more points for each additional player to give up. As always, patterns are harder to guess than you'd expect, so **whatever pattern you were thinking of, choose a simpler one!**

Make sure everyone gets an equal number of turns as designer. The highest overall score wins.

WIN, LOSE, BANANA

A GAME OF SOCIAL DEDUCTION

This epochal game is now out of print, and it only cost $1 anyway, so I don't feel bad giving away the rules. It's what game designer Marcus Ross calls "the minimal social deduction game," asking you to stare down your fellow players (hence "social") and extract the necessary information (hence "deduction"), ideally while having some fun (hence "game").

To play, you need **three players and three cards**, labeled "win," "lose," and "banana." To begin, deal one card facedown to each player. **Whoever receives "win" must guess which of the other two players is holding "banana."** Each of these players tries to convince the "win" player to pick them.

If the "win" player guesses right, then the victory is shared by "win" and "banana." If they guess wrong, then the victory goes to "lose."

FRANCO-PRUSSIAN LABYRINTH

A GAME OF STUMBLING IN THE DARK

To begin, draw two **9-by-9 grids**: one to track your own movements, and one to serve as a labyrinth for your opponent.

On the labyrinth, place **30 wall segments** anywhere you like, as long as they leave an **open path from the starting square (A1) to the ending square (I9).**

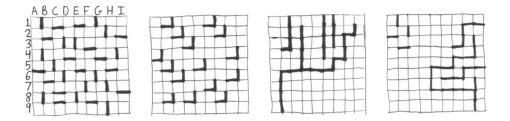

On each turn, **move one square at a time, in any direction**. After each attempted step, your opponent tells you whether you hit a wall. **Your turn ends after your fifth step, or when you hit a wall**, whichever occurs first. Your next turn begins in the square where you stopped.

Whoever reaches the bottom right corner first is the winner.

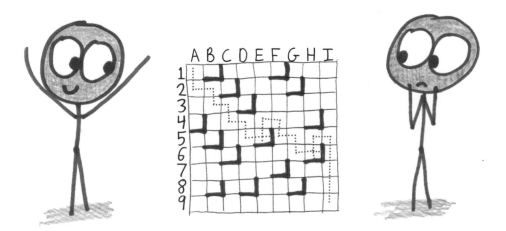

There is also a "French variation." It takes place on a 10-by-10 grid with 40 wall segments, and it follows a different rule for moving. On each turn, you pick a single direction, then travel in that direction until you are stopped (by a wall or the edge of the board). On your next turn, you may begin from any of the squares you occupied during your previous turn.

Andrea Angiolino suggests pretending that you are the Greek hero Theseus, hunting the beastly Minotaur through the mythic labyrinth. For a more modern alternative, pretend that you are lost in an Ikea. "There is no need to build a labyrinth," wrote Jorge Luis Borges, "when the entire world is one."

CONCLUSION

MY MOTHER ALWAYS enforced an ironclad rule: educational computer games only. My siblings and I grew up on Math Blaster, Yukon Trail, The Time Warp of Dr. Brain, and the Logical Journey of the Zoombinis. It was a good life—or at least, an academically nutritious one.

After my mother's death, our world fell into flux, and at age 13 I finally got my hands on the junk food I had long craved: an NHL hockey game.

My favorite feature was season mode. Pick a team, then play all 82 games, plus the playoffs. (To speed things up, you could have the computer simulate some of these games.) Season mode even let you trade players according to a simplistic algorithm: Every player was scored from 1 to 100, and opposing teams would consent to any trade that was roughly fair—say, a 67 for a 66, or an 82 for an 84.

That bit of wiggle room was crucial. It meant you could string together dozens of minor upgrades—68 to 70 to 71 to 73 to 74—and, with enough patience, eventually exchange a benchwarmer for an all-star. It was slow. It was rote. It amounted to toggling through endless menus. But it worked.

So what did I do with my prized game? I spent a few hours turning the Boston Bruins (my hometown team) into an unbeatable juggernaut, then simulated season after season, watching them smash records and win strings of championships, the greatest dynasty in sports history. I never played a game myself. I just presided fondly over this rigged universe, like a Greek god playing favorites. The league was a puzzle, and I had solved it.

Looking back, my mother needn't have worried about non-educational computer games. Her son could turn even the most pleasurable and frivolous of them into a spreadsheet.

What are you doing?   Playing hockey, obviously.

$$\sum_{i=1}^{n} r_{i,j} x_i < \sum_{i=1}^{n} r_{i,j+1} x_i$$

$$\sum_{i=1}^{n} r_{i,j} = \sum_{i=1}^{n} r_{i,j''}$$

where $x_i \in [0, 100]$,

$r_{i,j} = 1$ iff player i is on the Bruins at t_j

I plead guilty to killing NHL 2002. In my defense, that's what mathematicians have always done. Mathematicians love games the way a biology teacher loves frogs: with a love that is genuine but fatal. They will dissect the poor things limb from limb, master their inner workings, and then stand over the field of froggy corpses, declaring "Game solved!"

By "game" the mathematician means "puzzle," and by "solved," they mean "annihilated," reduced from an open-ended improvisation to a single, foregone outcome.

Ooh! You're beautiful! Show me your organs, will you?

I've tried to populate this book with games that are tricky to solve. My rationale is simple: Brains that never stop learning need systems that never stop teaching. Each game, at its best, is an inexhaustible collection of puzzles, created in close collaboration with your opponent. My move poses a puzzle for you. Your reply creates a fresh puzzle for me. And so on, and so on. From this vantage, chess is nothing but a generator of chess problems, a kind of perpetual puzzle machine.

Still, you may manage to solve some of the simpler games in this book. Therein lies the fundamental paradox of mathematical play: The joy of games is their insolvability, yet mathematicians keep insisting on solving them.

Early in writing this book, I read collections of Martin Gardner's classic *Scientific American* column, "Mathematical Games." At first, his selections confused me. Some of his games required unwieldy equipment. (Where do I buy a 30-by-30 chessboard, Martin?) Others existed in a haze of rule variations, with Gardner never specifying which ruleset he recommended. Worse still, almost all the games struck me as cold, Nim-like abstractions. Sure, they were ripe for mathematical analysis. But bust them out on family board game night, and they'd land like a stink bomb. Was this Gardner's idea of play?

Yes, actually. Though he called them "games," they were actually puzzles. You had to ascend to a higher plane to find the real game. It was the meta-game of devising new rules, the logic game of inventing new logic games. Gardner knew that mathematical play is not about coloring inside the lines, but about drawing new lines altogether. "Finite players play within boundaries," wrote James Carse. "Infinite players play with boundaries."

We're coming to the end of the book, so let's finish with the tale of a subtle rule change that transformed a whole game—a little adjustment to the boundary that wound up redefining the whole territory.

In 1994, soccer teams from 21 nations gathered to compete for the Caribbean Cup. All the standard rules were in place, with one twist in the tournament's tiebreakers: *If your team triumphed in sudden-death overtime, then the winning goal would count double.*

A tiny adjustment to the game's logic. Surely its effects would be just as tiny. Right?

On January 27, Barbados entered its match with Grenada needing a two-goal victory to advance on tiebreakers. They were on track, leading 2 to 0 until the 83rd minute, when a Grenada goal cut the lead to 2 to 1. Their spirits sank. Unless they could score in the final seven minutes, their tournament was over, and Grenada would advance in their place.

Then Barbados realized: What if the game went to overtime instead? A sudden-death victory would give them the two-goal win they required. Obeying the tournament's own logic, Barbados promptly scored on its own net, tying the game at 2 to 2.

Baffled at first, Grenada soon caught on, and tried to reverse the trick by scoring on *their* own goal. But Barbados was faster and defended the opposition net as if it were their own. That's when soccer's normal logic unraveled. Whereas Barbados needed to keep the score tied, Grenada was eager to break the tie in either direction. Thus, for five surreal minutes, fans watched slack-jawed as Grenada tried desperately to score at either end of the field, and Barbados staunchly defended both nets.

At last time expired. Barbados scored in overtime, securing a spot in the next round.

This is the power of mathematical play. Tweak the underlying logic, and you can turn a stadium of seasoned professionals into a backyard of scrambling kids, the goals of the game shifting moment by moment, as they chase the endless permutations of play.

VARIATIONS ON THIS BOOK, AND RELATED EXPERIENCES

BOARD GAMES: This book is about games you can play with materials you've got at home. But if you're looking for a more tactile experience, and willing to spend a few extra bucks, the tabletop gaming world offers delights ranging from casual party games to intense two-player abstracts. Here are a dozen to get you started:

Azul (2017, Next Move Games)

Blokus (2000, Educational Insights)

Prime Climb (2014, Math for Love)

Quarto (1991, Gigamic)

Qwirkle (2006, MindWare)

Root (2018, Leder Games)

Santorini (2016, Roxley)

SET (1998, SET Enterprises)

Tigris & Euphrates (1997, Hans im Gluck)

Wavelength (2019, Palm Court)

Wingspan (2019, Stonemaier Games)

Wits & Wagers (2005, North Star Games)

MATHEMATICAL PUZZLES: This is a book about unsolvable games, but solvable puzzles offer their own distinct pleasure. I recommend Martin Gardner (see the bibliography), Raymond Smullyan (e.g., the classic knight-knave puzzles in *What Is the Name of This Book?*), Alex Bellos (e.g., the eclectic fare of *Can You Solve My Problems?*), and Catriona Agg (whose celebrated geometry puzzles can be found on her Twitter account, @cshearer41). Other favorites of mine include *Geometric Snacks* by Ed Southall and Vincent Pantaloni, and *Area Mazes* by Naoki Inaba and Ryoichi Murakami.

MATHGAMESWITHBADDRAWINGS.COM: Here, you'll find online versions of several games, including Sequencium, Prophecies, and Dandelions, as well as printable grids, bonus games that didn't make the book, and more fun stuff. Thanks to my friend Adam Bildersee for his sterling work.

COMPREHENSIVE TABLES OF THE 75¼ GAMES IN THIS BOOK

"Which one is the quarter game?" you ask. Oh, friend, you don't know the half of it. The number 75¼ emerges from my innovative yet highly rigorous bookkeeping practices:

What Is It?	Book Value	Reasoning
A game with its own chapter (or its own section in a miscellany chapter)	1	That's what 1 means.
A game making a cameo appearance under "Variations and Related Games" in another game's chapter	$^{11}/_{12}$	Since these games are discussed only briefly, I offer them at a discount. You save 8.3% on each game.
A variation that differs meaningfully (but not completely) from the original, or a game that's actually a puzzle	$^{1}/_{4}$	I could fairly price these at ½ of a game each, but I prefer to offer an unbeatable bang for your buck/book.
A variation that still feels like the same game, in a vague yet fundamental way	$^{1}/_{57}$	There are 57 in the book, and I wanted the numbers to work out.

The following pages contain a complete inventory of this book's games, variations, varigames, and gamiations, organized by section and chapter. Within each chapter, all games require the same number of players and materials, except where noted.

	Section	Number of Games
I	Spatial Games	$14\ ^{9}/_{19}$
II	Number Games	$15\ ^{47}/_{114}$
III	Combination Games	$16\ ^{3}/_{19}$
IV	Games of Risk and Reward	$14\ ^{73}/_{76}$
V	Information Games	$14\ ^{14}/_{57}$
	Entire Book	$75\ ^{1}/_{4}$

All errors and inconsistencies are the fault of my friend Tom Burdett.[1]

---

1   "Why?" you ask. "Tom didn't make the tables." Exactly my point.

Spatial Games	Players	Materials	Tally
Dots and Boxes	2	paper, pen	1
Swedish Board			$^1/_{57}$
Dots and Triangles			$^1/_4$
Nazareno			$^1/_4$
Square Polyp		(*2 colors required*)	$^{11}/_{12}$
Sprouts	2+	paper, pen	1
Weeds			$^1/_4$
Point Set			$^1/_4$
Brussels Sprouts			$^1/_{57}$
Ultimate Tic-Tac-Toe	2	paper, pen	1
Single Victory			$^1/_{57}$
Majority Rules			$^1/_{57}$
Shared Territory			$^1/_{57}$
Ultimate Drop Three			$^1/_4$
The Dual Game			$^1/_4$
Dandelions	2	paper, pen	1
Balance Adjustments			$^1/_{57}$
Keeping Score			$^1/_{57}$
Random Plantings	(*1*)		$^1/_4$
Rival Dandelions		(*2 colors required*)	$^1/_4$
Collaborative			$^1/_{57}$
Quantum Tic-Tac-Toe	2	paper, pen	1
Many Worlds			$^1/_4$
Tournament Style			$^1/_4$
Quantum Chess		(*chessboard, chess set, coins*)	$^{11}/_{12}$
Miscellany			
Bunch of Grapes	2	paper, pen (2 colors)	1
Neutron	2	paper, pen, game pieces (5 of one kind, 5 of another, and 1 of a third kind)	1
Order and Chaos	2	paper, pen	1
Splatter	2	paper, pen (2 colors)	1
3D Tic-Tac-Toe	2	paper, pen	1
		ALL SPATIAL GAMES	14 $^9/_{19}$

Number Games	Players	Materials	Tally
Chopsticks	2+	hands	1
Chopsticks Mod N		*(paper, pens)*	$1/4$
Cutoff			$1/57$
Misère			$1/57$
One-Fingered Defeat			$1/57$
Suns			$1/57$
Zombies			$1/57$
Sequencium	2	paper, colored pens	1
Three Players	*(3)*		$1/57$
Four Players	*(4)*		$1/57$
Free Start			$1/57$
Fresh Seeds			$1/57$
Static Diagonals			$1/57$
33 to 99	2 to 5 or more	paper, pens, timer, 5 standard dice	1
The 24 Game		*(paper, pens, timer, four 10-sided dice)*	$11/12$
Banker		*(paper, pens, die)*	$11/12$
Number Boxes		*(paper, pens, one 10-sided or standard die)*	$11/12$
Pennywise	2 to 6	jar of coins	1
Other Starting Coinage			$1/57$
New Change-Making Rules			$1/4$
Flip	*(2)*	*(10 standard dice)*	$11/12$
Prophecies	2	paper, colored pens	1
Exotic Boards			$1/57$
Multiplayer	*(3 or 4)*		$1/57$
X-Prophecies		*(unsolved sudoku, colored pens)*	$1/57$
Sudoku Board			$1/4$
Miscellany			
Mediocrity	3, 5, or 7	paper, pen	1
Black Hole	2	paper, colored pens	1
Jam	2	paper, pen	$1/4$
Sir Boss's Barn *(in "Jam" section)*	2	paper, pen	$1/4$
Starlitaire	1	paper, pens (ideally colored)	$1/4$
Gridlock	2	graph paper, pens (ideally colored)	1

Tax Collector	1 or 2	paper, pen	1
Love and Marriage	15 to 50	paper, index cards, poster, pens	1
		ALL NUMBER GAMES	$15\,^{47}/_{114}$

Combination Games	Players	Materials	Tally
Sim	2	paper, colored pens	1
Whim Sim	*(3)*		$^1/_{57}$
Jim Sim			$^1/_{57}$
Lim Sim	*(6 to 30)*	*(poster, markers)*	$^1/_4$
Teeko	2	paper, 4 game pieces each of two kinds	1
Achi			$^{11}/_{12}$
All Queens Chess		*(6 game pieces each of two kinds)*	$^{11}/_{12}$
Teeko Classic			$^1/_{57}$
Neighbors	2 to 100	paper, pens, 10-sided die	1
Old-School Neighbors		*(paper, pens, deck of cards)*	$^1/_{57}$
Open Boards	*(2 to 8)*		$^1/_{57}$
Wordsworth	*(2 to 6)*	*(paper, pens)*	$^{11}/_{12}$
Corners	2	paper, colored pens	1
Multiplayer Corners	*(3 to 4)*		$^1/_{57}$
Perimeter-Style Corners			$^1/_{57}$
Quads and Quasars		*(large grid, jar of coins)*	$^{11}/_{12}$
Amazons	2	chessboard, jar of coins, 3 game pieces each of two kinds	1
6 by 6 Amazons		*(same, but only 2 game pieces each)*	$^1/_{57}$
10 by 10 Amazons		*(same, but need 4 game pieces each)*	$^1/_{57}$
Collector		*(paper, colored pens)*	$^{11}/_{12}$
Quadraphage		*(chessboard, jar of coins, any game piece)*	$^1/_4$
Pferdeäppel		*(chessboard, jar of coins, 2 knights)*	$^{11}/_{12}$

Miscellany			
Turning Points	2 or 4	grid, many pieces that can face specific direction (e.g., Goldfish)	1
Domineering	2	paper, pens (or dominos)	1
Hold That Line	2	paper, pen	1
Cats and Dogs	2	paper, pen	1
Row Call	2	paper, pen	1
		ALL COMBINATION GAMES	$16\,{}^{3}/_{19}$

Games of Risk and Reward	Players	Materials	Tally
Undercut	2	hands	1
Flaunt			$1/57$
Morra			$11/12$
Multiplayer Undercut	(3 to 4)		$1/57$
Underwhelm		(paper, pens)	$1/4$
Arpeggios	2	paper, pens, standard dice (2)	1
Multiplayer Arpeggios	(3 to 6)		$1/57$
Ascender	(1)		$1/4$
Ascenders	(2 to 10)		$11/12$
Outrangeous	3 to 8	paper, pens, internet access	1
Ratio Scoring			$1/57$
The Know-Nothing Trivia Game	(3)	(paper, pens)	$11/12$
Paper Boxing	2	paper, pens	1
Paper Boxing Classic			$1/57$
Paper Mixed Martial Arts			$1/57$
Blotto			$1/4$
Footsteps			$1/4$

Racetrack	2	graph paper, pens	1
Crash Penalties			$^1/_{57}$
Multiplayer Racetrack	(3 or 4)		$^1/_{57}$
Oil Spills			$^1/_{57}$
Point Grab			$^1/_{57}$
Slanted Start			$^1/_{57}$
Through the Gates			$^1/_{57}$
Miscellany			
Pig	2 to 8	paper, pens, 2 standard dice	1
Crossed	2	paper, colored pens	1
Rock, Paper, Scissors, Lizard, Spock	2	hands	1
101 and You're Done	2 to 4	paper, pens, 1 standard die	1
The Con Game	10 to 500	index cards (10 per player), pens	1
Breaking Rank	3+	paper, pens, internet access	1
		ALL GAMES OF RISK & REWARD	14 $^{73}/_{76}$

Information Games	Players	Materials	Tally
Bullseyes and Close Calls	2	paper, pens	1
Repeats Allowed			$^1/_{57}$
Self-Incrimination			$^1/_{57}$
Spot the Lie			$^1/_{57}$
Tight Lips			$^1/_{57}$
Jotto			$^{11}/_{12}$
Caveat Emptor	2 to 8	paper, pens, 5 household objects	1
Caveat Emptor with Real Auctions			$^1/_{57}$
Liar's Dice		(per player: 5 standard dice, 1 cup)	$^{11}/_{12}$
Liar's Poker		(per player: one dollar bill)	$^1/_4$

LAP	2	paper, pens (ideally 4 colors)	1
Beginners' LAP			$^{1}/_{57}$
Experts' LAP			$^{1}/_{57}$
Classic LAP			$^{1}/_{57}$
Rainbow Logic			$^{1}/_{57}$
Quantum Go Fish	3 to 8	paper, pens, paper clips (*or, if you're daring, just hands*)	1
Lose a Turn			$^{1}/_{57}$
Playing On			$^{1}/_{57}$
Blind Kwartet			$^{1}/_{57}$
Saesara	3 to 5	paper, different colored pens	1
Speed Saesara			$^{1}/_{57}$
Grand Saesara			$^{1}/_{57}$
Jewels in the Sand	(*2 to 8*)	(*nothing*)	$^{11}/_{12}$
Miscellany			
Battleship	2	graph paper, pens	1
Quantum Hangman	2 to 10	paper, pens	1
Buried Treasure	2	graph paper, pens	1
Patterns II	3 to 5	graph paper, pens	1
Win, Lose, Banana	3	paper, pens	1
Franco-Prussian Labyrinth	2	graph paper, pens	1
		ALL INFORMATION GAMES	14 $^{14}/_{57}$

CLOSING CREDITS

I'm writing this in May 2021, a few hours after receiving my final dose of the COVID vaccine. It's a bittersweet day: As I say goodbye to this awful, misshapen year, I must also say goodbye to the project that got me through it. To the people who helped this book along, my debt of gratitude is tremendous, well above 100% of GDP. By enriching the book, you enriched my life, too.

CAST (IN APPROXIMATE ORDER OF APPEARANCE)

Editor (*who proposed this book concept in February 2019, and gently persuaded me not to tell rambling 3,500-word stories like some kind of millennial grandpa*): Becky Koh

Consiglieri (*who got me into this whole book-writing career, and whose presence in my life makes me pity the poor folks with just one literary agent*): Dado Derviskadic, Steve Troha

Math and Gaming Pals (*who, before the pandemic descended upon us, provided wonderful feedback and even better company*): Abby Marsh, Joe Rosenthal, Matt Donald, Phil McDonald, Taylor McDonald, Brackett Robertson, Andrew Roy, Jeffrey Bye, Rob Liebhart, Vito Sauro, Geoff Koslig, Mary Koslig, Jim Orlin, Jenna Laib, Cash Orlin, Justin Palermo, Denise Gaskins, John Golden, Gord Hamilton, Dan Finkel, Andrew Beveridge, Dan O'Loughlin (whose book I intend to have returned by the time he reads this), Nathalie Vega-Rhodes, Jim Propp, Adam Bildersee, and the students and teachers of Saint Paul Academy

Game Designers (*who graciously welcomed a clumsy newcomer into their community and offered brilliant advice, some of which I was even smart enough to follow*): Everyone at Protospiel Minnesota 2020, along with Andy Juell and Joe Kisenwether, whose creations I'm proud to feature in this book

My Tribe of Play Testers (*who, when live events became impossible, graciously volunteered to try games via email; whose 1,300 surveys, on more than 40 different games, from more than 300 people, helped make this book possible; without whom I could not have done this at all; toward whom I feel no small measure of guilt, for failing to adequately reciprocate the patience,* *generosity, and selfless good cheer that they contributed to this project; of whom I only have space to list a handful; and from whom I will single out those who filled out the most surveys, as a meaningful if slightly arbitrary metric)*: Mihai Maruseac, Dylan Kane, Joseph Kisenwether, Jamie Roberts, Katie McDermott (whose generous, honest, life-affirming surveys were always a delight to read, and helped me remember why I wanted to do this book), Andy Juell, Zack, F.J. Plummer, Yann Jeanrenaud, Shriya Navil, Lisa, Eric Haines, Filomena Joana, Kim, Connie Barnes, Glen Lim, Julie Bellingham, Isabel Anderson, Scott Mittman, Roxanne Pittard, Steven Lundy, Michelle Ciccolo, Immanuel Balete, Paul Fonstad, John Haslegrave, Flo, Malachi Kutner, Michelle Celich, Steven Goldman, Kelly Burke, Marina Shrago, Sara Jensen (who suggested the great title "Parents vs. Kids" for Order and Chaos), Nathaniel Ou, Guillaume Douville, Denise, Stephanie Moore, Paolo Imori, Katrijn, Mona Hennigar, Emily Dennett, Aaron Carpenter, Debby Vivari, Cory, Carole Bilyk, Jessie Oehrlein, William Kho, Tim Newton, Valkhiya, Cindy Falla, Anastasia Martin, Shira, Eric Hanson, Michal Rudolf, Rich Beverungen, Shannon Jeter, Norma Gordon, and Archita. Additional shout-out to the folks who gave helpful feedback on how the games might appear to colorblind readers: Thom Fries, Shriya Navil and family, and Christian Lawson-Perfect.

Web Maestro (*who built the gorgeous site MathGamesWithBadDrawings .com, which you should definitely check out*): Adam Bildersee

Producers (*who transformed a misshapen bundle of oddly formatted documents into a glorious Real Book*): Kara Thornton, Betsy Hulsebosch, Hannah Jones, Melanie Gold, Katie Benezra, Paul Kepple, Alex Bruce, Lori Paximadis, Francesca Begos, and everyone at Black Dog & Leventhal

I know I've omitted countless names. Sorry to those names, and to the people they represent. Special thanks to my family, my friends, my colleagues, my students, my Taryn, and my Casey.

That's the end. But if you want to keep staring at this page, in hopes of seeing Josh Brolin put on a fancy glove (or whatever it is we're waiting for at the end of those Marvel films—I never know), then feel free to stick around.

HEY, WHY IS YOUR BIBLIOGRAPHY ORGANIZED AS A LIST OF "FREQUENTLY ASKED QUESTIONS"?

Because it's my book, friend, and I can make weird structural choices if I want.

Are these actual questions? Of course. They have question marks and everything.

No, I mean, is it true that they're "frequently asked"? That depends on your definitions of "frequently" and "asked."

I see. So these "questions" are just thin pretexts for mentioning the sources you used. How dare you. That's an accusation, not a question. Also, yes.

All right then. What books did you use? I'm so glad you asked! Setting aside books I used for specific chapters (which I'll mention later), here are the ones I used for general research:

Michael Albert, Richard Nowakowski, and David Wolfe, *Lessons in Play: An Introduction to Combinatorial Game Theory* (Wellesley, MA: A. K. Peters, 2007).

Leigh Anderson, *The Games Bible: Over 300 Games: The Rules, the Gear, the Strategies* (New York: Workman Publishing Company, 2010).

Andrea Angiolino, *Super Sharp Pencil and Paper Games* (New York: Sterling, 1995). Source for Franco-Prussian Labyrinth, Racetrack (as "The Track"), Dots and Triangles (as "Triangles"), Nazareno, Footsteps (as "Battle of the Stars"), and Bullseyes & Close Calls (as "Little Numbers").

R. C. Bell, *Board and Table Games from Many Civilizations* (New York: Dover Publications, 1979).

Elwyn Berlekamp, John Conway, and Richard Guy, *Winning Ways for Your Mathematical Plays, Volumes 1, 2, 3, and 4* (Natick, MA: A. K. Peters, 2001). Source for Domineering, Dots and Boxes, Sprouts, Jam, and Sir Boss's Barn (as "Hot").

Roger Caillois, *Man, Play, and Games*, trans. Meyer Barash (Champaign: University of Illinois Press, 2001).

James Carse, *Finite and Infinite Games* (New York: Free Press, 1986).

Greg Costikyan, *Uncertainty in Games* (Cambridge, MA: MIT Press, 2013).

James Ernest, *Chief Herman's Holiday Fun Pack: Instruction Booklet and Guide to Better Living* (Seattle: Cheapass Games, 2000). Source for Pennywise, Flip, the Con Game, and Love and Marriage.

Skip Frey, *Complete Book of Dice Games* (New York: Hart Pub. Co, 1975). Source for Pig.

Martin Gardner, *Knotted Doughnuts and Other Mathematical Entertainments* (New York: W. H. Freeman, 1986). Source for Sim, Racetrack, Quadraphage.

Martin Gardner, *Mathematical Carnival* (New York: Penguin, 1990). Source for Sprouts, Jam, and Sir Boss's Barn (as "Hot"). Gardner stalks the pages of this book, which might very well not exist without the legacy of his "Mathematical Games" column in *Scientific American*. His contributions go deeper than math games: Douglas Hofstadter called him "one of the greatest intellects produced in this country in this century," and Stephen Jay Gould dubbed him "the single brightest beacon defending rationality and good science against the mysticism and anti-intellectualism that surround us." I believe his most underrated virtue was his taste; no one else has quite his knack for problems of (a) broad accessibility and (b) real intellectual depth.

Douglas Hofstadter, *Metamagical Themas* (New York: Basic Books, 1985). Source for Undercut, Flaunt, Underwhelm, Mediocrity.

Johan Huizinga, *Homo Ludens: A Study of the Play-Element in Culture* (London: Redwood Burn Ltd, 1980).

Walter Joris, *100 Strategic Games for Pen and Paper* (London: Carlton Books, 2002). Source for Square Polyp, Point Set, Bunch of Grapes, Black Hole, and Collector.

Reiner Knizia, *Dice Games Properly Explained* (Blue Terrier Press). Source for 33 to 99 (as "Ninety Nine") and Banker.

David McAdams, *Game-Changer: Game Theory and the Art of Transforming Strategic Situations* (New York: W. W. Norton & Company, 2014).

Ivan Moscovich, *1,000 Playthinks: Puzzles, Paradoxes, Illusions, and Games* (New York: Workman Publishing Company, 2001). Source for Crossed (as "Playthink #216").

João Pedro Neto and Jorge Nuno Silva, *Mathematical Games: Abstract Games* (Mineola, NY: Dover 2013).

Oriol Ripoll, *Play with Us: 100 Games from Around the World* (Chicago: Chicago Review Press, 2005). Source for Morra.

Sid Sackson, *A Gamut of Games* (New York: Dover, 1969). Source for Hold That Line, Paper Boxing, LAP, and Patterns II.

R. Wayne Schmittberger, *New Rules for Classic Games* (New York: John Wiley & Sons, 1992). Source for Wordsworth (as "Crosswords") and for several variations, especially to Bullseyes and Close Calls.

John Sharp and David Thomas, *Fun, Taste, and Games: An Aesthetics of the Idle, Unproductive, and Otherwise Playful* (Cambridge, MA: MIT Press, 2019).

Eric Solomon, *Games with Pencil and Paper* (Toronto: General Publishing Company, 1993). Source for Wordsworth (as "Think of a Letter"), Buried Treasure, and the pencil-and-paper version of Eleusis that inspired Saesara.

Francis Su, *Mathematics for Human Flourishing* (New Haven, CT: Yale University Press, 2020).

Brian Upton, *The Aesthetic of Play* (Cambridge, MA: MIT Press, 2015).

So, it was all books? No websites? Hardly! Here's a small sampling:

The American Journal of Play is published in Rochester, New York, by the Strong National Museum of Play. https://www.journalofplay.org

BoardGameGeek.com is an essential resource and, more than that, a sprawling civilization full of game lovers who are thoughtful, helpful, enthusiastic, and exceedingly hard to please. I wouldn't have it any other way. http://boardgamegeek.com

Bona Ludo is a well-written blog on board games. http://bonaludo.com

Let's Play Math, run by Denise Gaskins, is a trove for teachers and homeschooling families. http://denisegaskins.com

Math for Love, run by Dan Finkel and Katherine Cook, features lots of great ideas for the classroom, as well as information about Prime Climb and Tiny Polka Dots, two award-winning (and, if I may say, totally excellent) mathematical board games. https://mathforlove.com

Math Hombre, the site of Professor John Golden, is full of great ideas and winning personality. http://mathhombre.blogspot.com

Math Pickle is another trove of gaming resources for the home and the classroom. Run by Lora Saarnio and Gord Hamilton, who designed the transcendently elegant board game Santorini. https://mathpickle.com

My Kind of Meeple, by Emily Sargeantson, is a cheerful and thoughtful blog on the world of tabletop games. https://mykindofmeeple.com

So Very Wrong about Games was among my favorites of the many tabletop gaming podcasts I explored (http://twitter.com/sowronggames). Others included *Breaking Into Board Games*, *Ludology*, and *This Game Is Broken*.

Talking Math with Your Kids is Christopher Danielson's beautiful project aiming to inspire mathematical conversations between grown-ups and children. His books *Which One Doesn't Belong?* and *How Many?* are great starting points, too. https://talkingmathwithkids.com

INTRODUCTION

Are you calling me a baby chimpanzee? Yes. Stephen Jay Gould said it first, and better. I recommend "A Biological Homage to Mickey Mouse" in *The Panda's Thumb: More Reflections in Natural History* (New York: W. W. Norton, 1980).

You went for simple games, but what's the most complicated board game ever? It's probably Campaign for North Africa. The rulebook, written in size 8 font, runs 90 pages. A single game requires 10 players and is said to last 1,500 hours. (That's an estimate; no completed game has ever been verified.) The game is a simulation as complicated as the reality it simulates, like Lewis Carroll's fictional map on the scale of a mile to the mile. Want a taste of the complexity? "Every game turn," one player told *Kotaku*, "three percent of the fuel evaporates, unless you're the British before a certain date, because they used 50-gallon drums instead of jerry cans. So instead, seven percent of their fuel evaporates" (Luke Winkie, "The Notorious Board Game That Takes 1,500 Hours to Complete," *Kotaku*, February 5, 2018, https://kotaku.com/the-notorious-board-game-that-takes-1500-hours-to-compl-1818510912). Having fun yet?

Where can I learn more about the making of Set? Danielle Steinberg, "Canine Epilepsy and Purple Squiggles: The Unexpected Success Story of SET," *Gizmodo*, August 23, 2018, https://gizmodo.com/canine-epilepsy-and-purple-squiggles-the-unexpected-su-1828527912.

Where can I learn more about the making of the Rubik's Cube? "The Perplexing Life of Erno Rubik," *Discover* 8, no. 8 (1986): 81. Full Text © Family Media Inc., 1986. URL: http://www.puzzlesolver.com/puzzle.php?id=29;page=15.

Did probability theory really begin from a gambling puzzle? History is never that simple, but it was certainly a pivotal moment. See more in Keith Devlin, *The Unfinished Game: Pascal, Fermat, and the Seventeenth-Century Letter That Made the World Modern* (New York: Basic Books, 2010).

Where can I learn more about the Bridges of Konigsberg? I learned the names of the bridges from Teo Paoletti, "Leonard Euler's Solution to the Konigsberg Bridge Problem," an undergraduate paper written for Professor Judit Kardos's History of Mathematics course at the College of New Jersey, https://www.maa.org/press/periodicals/convergence/leonard-eulers-solution-to-the-konigsberg-bridge-problem.

Where can I learn more about John Conway? I quote from Jim Propp's memorial post, "Confessions of a Conway Groupie," *Mathematical Enchantments*, May 16, 2020, https://mathenchant.wordpress.com/2020/05/16/confessions-of-a-conway-groupie. I also recommend the obituary written by his biographer: Siobhan Roberts, "John Horton Conway, a 'Magical Genius' in Math, Dies at 82," *New York Times*, April 15, 2020.

I. SPATIAL GAMES

INTRODUCTION

When I play Asteroids, am I really flying inside a giant doughnut? Not at all. You're flying *across the surface* of a giant doughnut. For more, check out the episode "Klein Bottle with Matthew Scroggs" from the *Mathematical Objects* podcast, hosted by Katie Steckles and Peter Rowlett.

Where did Ingrid Daubechies say that thing about the geometry of doll clothing? I got the quote from Denise Gaskins's excellent list "Math and Education Quotations." It's found at https://denisegaskins.com/best-of-the-blog/quotations. She drew from J. J. O'Connor and E. F. Robertson, "Ingrid Daubechies," *MacTutor History of Mathematics* (University of St Andrew, Scotland, September 2013), https://mathshistory.st-andrews.ac.uk/Biographies/Daubechies.

Who's this M. C. Escher character? He's every mathematician's favorite visual artist. The quotes come from *Escher on Escher: Exploring the Infinite* (New York: Harry N. Abrams, 1989).

Who was this "Henri Poincaré" who called geometry not "true" but merely "convenient"? Some kind of disgruntled student? Poincaré is sometimes described as the last mathematician who mastered all of his era's mathematics, so "disgruntled student" is fair enough. I got the quote from page 50 of his book *Science and Hypothesis* (New York: Dover Publications, 1952).

Is this mathematician John Urschel the same guy as former NFL player John Urschel? Yup. Known to his teammates as Ursch. The quote is from his memoir, written with Louisa Thomas, *Mind and Matter: A Life in Math and Football* (New York: Penguin Press, 2019).

DOTS AND BOXES

What's the deal with this game? Its first publication was Edouard Lucas, *L'Arithmetique Amusante: Introduction Aux Recreations Mathematiques* (Paris: Gauthier-Villars et Fils Imprimeurs-Libraires, 1895). It's on Google Books.

No, I mean, how do I win? If you're serious about strategy, check out Elwyn Berlekamp's definitive book *The Dots and Boxes Game: Sophisticated Child's Play* (Oxfordshire: Routledge, 2000). In an email to me, computer graphics expert Eric Haines recalled meeting Berlekamp at a conference, where he was "playing all comers in Dots & Boxes, crushing us like bugs." There's also a detailed discussion in *Winning Ways for Your Mathematical Plays*. Or, if you dwell on the interwebs, Canadian mathematician Ilan Vardi wrote up a nice site "Mathter of the Game," originally at GeoCities, now salvaged at http://www.chronomaitre.org/dots.html. Also check out Julian West, "Championship-Level Play of Dots-and-Boxes," *Games of No Chance*, MSRI Publications 29, 1996.

Why does this game go by so many names? Just as the First Peoples of Canada need lots of words for snow, we postindustrial peoples need lots of words for our bourgeois leisure activities. Anyway, I appreciate folks on Twitter for sharing the various international names for the game. Hat tips to @misterwootube, @OlafDoschke, @mathforge, @ConorJTobin, @marioalberto, @LudwigBald, @LauraKinnel, and @relinde, among others.

Did you come up with those names for the Square Polyp shapes? Or did Walter Joris? Actually, it was Joe Kisenwether and his pals.

SPROUTS

Why is "curious topological flavor" in quotation marks? Martin Gardner introduced the world to Sprouts by quoting a letter from mathematics student David Hartshorne. "A friend of mine," David wrote, "a classics student at Cambridge, introduced me recently to a game called 'Sprouts' which became a craze at Cambridge last term. The game has a curious topological flavor." Source: Gardner, *Mathematical Carnival*.

Wow, topology is so cool! Isn't it? I eagerly recommend David Richeson, *Euler's Gem: The Polyhedron Formula and the Birth of Topology* (Princeton, NJ: Princeton University Press, 2012).

This John Conway fella sounds like quite a character. You can say that again. For a delightful biography, read Siobhan Roberts, *Genius at Play: The Curious Mind of John Horton Conway* (New York: Bloomsbury USA, 2015).

So how do you win at Sprouts? For strategic discussion, you still can't beat *Winning Ways for Your Mathematical Plays*. That said, computers have now overtaken humans, and then some. Learn more from Julien Lemoine and Simon Viennot, "Computer Analysis of Sprouts with Nimbers," *Games of No Chance 4*, MSRI Publications 63, 2015.

Uh, what if I don't want to read the academic literature? My own research was accelerated and enriched by a YouTube video from Kevin Lieber on the channel Vsauce2. Search for "The Dot Game That Breaks Your Brain."

What if the game goes on forever? Can't happen. The game beginning with n spots can last at most $3n - 1$ moves. Here's a quick proof showing why. First, note that each spot has three "lives." Thus, an n-spot game begins with $3n$ lives. Each move consumes two lives and introduces a new one, decreasing the total number of lives by one. Since the game cannot continue with only one life left, it must stop on or before the $(3n - 1)$th move.

ULTIMATE TIC-TAC-TOE

How can I learn more about fractals? For a deep dive, cowritten by a fractal geometer and a poet (has any book had a more perfect author combo?), read Michael Frame and Amelia Urry, *Fractal Worlds: Grown, Built, and Imagined* (New Haven, CT: Yale University Press, 2016).

What if I just want to look at pretty nature pictures? Try *Nature's Chaos* (New York: Little, Brown, 2001), with lyrical writing by James Gleick (author of *Chaos*) accompanying photos of natural fractals by Eliot Porter. Or you can search for Paul Bourke's site Google Earth Fractals.

When you say Trump was "playing Ultimate Tic-Tac-Toe," do you mean that in a good way or a bad way? As a firm believer that political persuasion is best accomplished through the endnotes of cartoon-illustrated books, I'm tempted to share my spicy takes here. But I'm not actually the one who came up with the analogy. It's from Oliver Roeder, "Trump Isn't Playing 3D Chess—He's Playing Ultimate Tic-Tac-Toe," *FiveThirtyEight*, May 7, 2018.

Did Plato really say the whole world is made of special right triangles? Yeah, it's in *Timaeus*. He claims that everything is made of fire, earth, water, and air; that these are "bodies"; and that "every sort of body possesses solidity, and every solid must necessarily be contained in planes; and every plane rectilinear figure is composed of triangles; and all triangles are originally of two kinds . . . " Pretty hard to argue with that, insofar as it's hard to argue with a hot bowl of nonsense.

Where did you get that Robert Frost quote? I can't find it online. Oh, I made it up. Sounds much better coming from Frost, doesn't it?

What?! That's an egregious breach of intellectual honesty! Hey, it's not half as bad as how everyone quotes "The Road Not Taken" as if it's triumphant life wisdom ("I took the one less traveled by") instead of a sly comment on how hindsight assigns false grandeur to arbitrary decisions ("the passing there / had worn them really about the same").

DANDELIONS

What sources did you use for this chapter? None. This game burst from my mind like Athena from the forehead of Zeus, full-grown and ready for battle. I relinquish the glory to no one.

What about the dozen or so people you mention in the chapter? Oh yeah, they had great suggestions and insights. I relinquish *some* glory to them.

And the dozens of others who helped to play-test the game? Oh, right, them, too. Thanks, everybody. You're the best. But other than that: no one!

QUANTUM TIC-TAC-TOE

How good a metaphor for quantum mechanics is this game? Pretty good. In fact, its creator presents it as a teaching tool. Check out Allan Goff, "Quantum Tic-Tac-Toe: A Teaching Metaphor for Superposition in Quantum Mechanics," *American Journal of Physics* 74, no. 11 (2006), as well as Allan Goff, Dale Lehmann, and Joel Siegel, "Quantum Tic-Tac-Toe, Spooky-Coins and Magic-Envelopes, as Metaphors for Relativistic Quantum Physics," paper presented at the AIAA/ASME/SAE/ASEE Joint Propulsion Conference and Exhibit, 2002, https://doi.org/10.2514/6.2002-3763.

I'm kind of afraid to ask, but how can I learn more quantum physics? The two books on my shelf are Chad Orzel, *How to Teach Quantum Physics to Your Dog* (New York: Scribner, 2010), and Philip Ball, *Beyond Weird: Why Everything You Thought You Knew about Quantum Physics Is Different* (Chicago: University of Chicago Press, 2020).

Wouldn't it be more "quantum" if the game had shifting rules? Funny you should ask. Goff has written that the game's rules felt almost inevitable, but in fact, the same premise has occurred to other people and resulted in quite different rulesets. For example, the app Quantum TiqTaqToe (https://quantumfrontiers.com/2019/07/15/tiqtaqtoe) has a gradual buildup of quantum features. I recommend it. Or, for a more scholarly approach to the game, see J. N. Leaw and S. A. Cheong, "Strategic Insights from Playing the Quantum Tic-Tac-Toe," posted to arXiv.org in 2010 (https://arxiv.org/pdf/1007.3601.pdf).

II. NUMBER GAMES

INTRODUCTION

If every number is interesting, then what about [*fill in number here*]? For a constructive proof that every number is indeed interesting, check out David Wells, *The Penguin Dictionary of Curious and Interesting Numbers* (New York: Penguin Books, 1998). Or, for a lovely and uplifting corollary of the "every number is interesting" proof, check out Susan D'Agostino, "Every Minute of Your Life Has Been Interesting," *Journal of Humanistic Mathematics* 7, no. 1 (2017): 117–118.

Where can I learn more about aliquot sequences? Go seek the eighth wonder of the world: the On-Line Encyclopedia of Integer Sequences (http://oeis.org). You can find just about everything there, including perfect numbers (A000396), amicable numbers (A259180), and that ridiculous cycle of 28 sociable numbers (A072890).

Did your friend Julian really say that pure math "keeps mathematicians off the streets"? Yes. Date: 2003. Source: I was there. However, I wasn't there for John Littlewood's quip that perfect numbers have done "no good, and no harm, either"; it comes from John Littlewood, *A Mathematician's Miscellany* (London: Methuen & Co Ltd, 1953).

CHOPSTICKS

Where can I learn more about this game, such as how to crush my opponents? The children of the world have posted dozens of instructional videos on YouTube, some of which show guaranteed wins for certain rule sets, and all of which are adorable. Meanwhile, on Wikipedia, you'll find a solid mathematical analysis and a comprehensive list of variations.

Whoa there, Mr. Orlin. Did you seriously research a whole chapter of your published book using only YouTube and Wikipedia? Um . . . do as I say, students, not as I do!

SEQUENCIUM

Does White really have a big advantage in chess? Yes. At the highest levels, it's assumed that White plays for the win, and Black for the draw. The chapter's chess quotes come from Gary Alan Fine, *Players and Pawns: How Chess Builds Community and Culture* (Chicago: University of Chicago Press, 2015).

Who came up with that wild Thue-Morse sequence? Well, it's named after Axel Thue and Marston Morse, so, of course, it was first studied by Eugene Prouhet. I first learned it from a talk by Phil Harvey titled "Cumulative Fairness," delivered at the UK's MathsJam Conference, November 7–8, 2015.

What kind of silly applications have mathematicians come up with for Thue-Morse? Here's a partial list: Marc Abrahams, "How to Pour the Perfect Cup of Coffee," *Guardian*, July 12, 2010; Joshua Cooper and Aaron Dutle, "Greedy Galois Games," https://people.math.sc.edu/cooper /ThueMorseDueling.pdf; Ignacio Palacios-Huerta, "Tournaments, Fairness, and the Prouhet-Thue-Morse Sequence," *Economic Inquiry* 50, no. 3 (2012): 848–849; and my personal favorite, Lionel Levin and Katherine E. Stange, "How to Make the Most of a Shared Meal: Plan the Last Bite First," *American Mathematical Monthly* 119, no. 7 (2012): 550–565.

33 TO 99

Where can I watch that Japanese commercial? Just search YouTube for "Nexus 7: 10 Puzzle." I learned about it from Gary Antonick, "Can You Crack the 24 Puzzle, and the 10 Puzzle That Went Viral in Japan?" *New York Times*, September 7, 2015.

Hey, isn't this game just "the 24 puzzle"? Guilty as charged. Still, it really spices things up to change the target number each round. For a bit of history on the 24 Game, check out John McLeod, "Twenty-Four," https://www.pagat.com/adders/24.html. There's also a simple and pleasing online version of the 24 game at http://4nums.com.

And isn't it also just like the "Four Fours" problem? Indeed it is. Pat Ballew has a good history at "Before There Were Four-Fours, There Were Four Threes, and Several Others," *Pat'sBlog*, December 30, 2018, https://pballew.blogspot.com/2018/12/before-there-were-four-fours-there-were.html. And if you need spoilers, Paul Bourke has compiled an impressive list of solutions at http://paulbourke.net/fun/4444.

How many different numbers can you make with five dice? It depends what you mean. Are you counting fractions? Negative numbers? What about the same number created multiple different ways? Anyway, here are some results for a few sets of dice.

Dice	How Many Values Can You Create?	How Many of These Are Whole Numbers?	How Many of the Values from 33 to 99 Are Achievable?
1, 2, 3, 4, 5	3,068	117	60 out of 67
2, 3, 4, 5, 6	5,281	222	61 out of 67
2, 2, 3, 3, 5	1,722	81	45 out of 67
4, 4, 4, 4, 4	200	35	13 out of 67
1, 2, 3, 4, 7	4,027	150	67 out of 67

I'm not British. What is this *Countdown* business? YouTube has countless *Countdown* videos of Rachel Riley working her magic. If you want the one where she hits 649 on the nose, it's here: https://youtu.be/9eMs_o08Gm4?t=295.

Hey, I'm a teacher. Tell me more about this "Number Boxes" stuff. Okay, non-teachers, scram! This paragraph is not for you. [*waits*] All right, now that it's just us educators: I attribute that version to Marilyn Burns, "4 Win-Win Math Games," *Do the Math*, March/April 2009. The math education site nRich also has a nice discussion (https://nrich.maths.org/6606), as does educator Jenna Laib, "One of My Favorite Games: Number Boxes," *Embrace the Challenge*, May 29, 2019 (https://jennalaib.wordpress.com/2019/05/29/one-of-my-favorite-games-number-boxes). In a similar vein, I give a strong recommendation to Open Middle problems (http://openmiddle.com), created by Nanette Johnson and Robert Kaplinsky.

PENNYWISE

Where does this game come from? The inimitable James Ernest of Cheapass Games (http:// cheapass.com). This is a family book, so I hoped to avoid printing the word "ass" too many times, but James forced my hand. Also, my ass.

Did human writing really come from sheep tokens? Arguably, yes. To dive into the anthropology, read Denise Schmandt-Besserat, "Tokens: Their Significance for the Origins of Counting and Writing" (https://sites.utexas.edu/dsb/tokens/tokens/). Then, for more depth, see Denise Schmandt-Besserat, "Two Precursors of Writing: Plain and Complex Tokens," in *The Origins of Writing*, edited by Wayne M. Senner (Lincoln: University of Nebraska Press, 1991), 27–41.

What four denominations allow you to make every value from 1¢ to 99¢ with the fewest total coins? The denominations you're looking for are 1¢, 5¢, 18¢, and 25¢. Together, they can make every value from 1¢ to 99¢ using a total of just 389 coins, which is the fewest possible.

I find it surprising that this optimal system has three coins in common with our existing system. And by "surprising," I mean "boring." My personal favorite has the denominations 1¢, 3¢, 13¢, and 31¢. It requires a total of 400 coins to make all that change, but I say it's worth it for the anarchy of 13¢ and 31¢ coins.

Sidebar: We usually make change by starting with the biggest coins and moving down. For example, to make 72¢, we use as many quarters as possible (2), then as many dimes as possible (2), then as many nickels as possible (0), then as many pennies as necessary (2). This is called the "greedy algorithm," and in our penny-nickel-dime-quarter system, it minimizes the number of coins.

But that's not true in the penny-nickel-18¢-quarter system. For example, the greedy algorithm would tell you to make 72¢ by using seven coins (two quarters, one 18¢, and four pennies), when there's an option using just four coins (all of them 18¢ coins). So in this world, making efficient change is a lot trickier!

If you insist on using only the greedy algorithm, then the best system is 1¢, 3¢, 11¢, and 37¢ (requiring 410 coins).

Could either of the new change-making rules lead to an endless game? Nope. Let's consider "perfect change" first. Each penny is still worth one turn. Each nickel is worth up to six turns (one turn to exchange it for pennies, then five penny turns). Each dime is worth up to 13 turns (one to exchange it for nickels, then six per nickel). And each quarter is worth up to 33 turns (one to exchange it for two dimes and a nickel, then 13 per dime and six per nickel). So your original change is worth at most $33 + (13 + 13) + (6 + 6 + 6 + 6) + (1 + 1 + 1 + 1 + 1) = 88$ turns.

What about the "more than perfect change" version? Let's say you have five players. A penny is still worth one turn. A nickel, at best, can be exchanged for all 20 pennies in the game, and is thus worth at most 21 turns. A dime, at best, can be exchanged for all 15 nickels (worth at most 21 turns each) and all 20 pennies (worth one turn each), for a total of 336 turns. And a quarter, at best, can be exchanged for all 10 dimes (worth at most 336 turns each), plus all the nickels and quarters (which are equivalent to an 11th dime), for a total of 3,696 turns. Thus, you absolutely cannot last more than 4,435 turns.

PROPHECIES

Whoa. This self-referential stuff blows my mind. That's only the beginning; your mind has lots of material left to combust. To make it go boom, dig into the work of Douglas Hofstadter, author of *Gödel Escher Bach: An Eternal Golden Braid* (New York: Basic Books, 1979) and *Metamagical Themas: Questing for the Essence of Mind and Pattern* (New York: Basic Books, 1985). Or, for more on Bertrand Russell, Kurt Gödel, and the history of 20th-century logic, I recommend Apostolos Doxiadis and Christos Papadimitriou, *Logicomix: An Epic Search for Truth* (New York: Bloomsbury USA, 1999).

I don't want to just read; I want to do some problems! Well, for a fascinating (and tricky) series of birdcall-themed puzzles, leading from elementary logic all the way to the brink of Gödel's theorems, I recommend the irreplaceable Raymond Smullyan, *To Mock a Mockingbird* (New York: Oxford University Press, 1982).

Nah, I don't want to read a whole book. Just give me a taste. Okay, here's a roster of self-referential sentences from Hofstadter's *Metamagical Themas*. They range from gentle brain ticklers to merciless brain punchers:

This sentence no verb.

This sentence contains exactly threee erors.

As long as you are not reading this sentence, its fourth word has no referent.

Thit sentence is not self-referential because "thit" is not a word.

If I had finished this sentence,

This sentence is not about itself, but about whether it is about itself.

I have nothing to say, and I am saying it.

This sentence does in fact not have the property it claims not to have.

How did you come up with that self-referential table? It's a classic puzzle, though I haven't seen it done with tables before. See Alex Bogomolny, "Place Value," *Cut the Knot!* July 1999 (https://www.cut-the-knot.org/ctk/SelfDescriptive.shtml).

What are the other solutions for the self-descriptive table? Here they are, including the one in the chapter.

Digit	1	2	3	4
Appearances	2	3	2	1

Digit	1	2	3	4
Appearances	3	1	3	1

Digit	1	2	3	4	5
Appearances	3	2	3	1	1

Digit	1	2	3	4	5	6	7
Appearances	4	3	2	2	1	1	1

Digit	1	2	3	4	5	6	7	8
Appearances	5	3	2	1	2	1	1	1

Digit	1	2	3	4	5	6	7	8	9
Appearances	6	3	2	1	1	2	1	1	1

If you want your table to include the digits from 1 to n, that's an exhaustive list.

OTHER NUMBER GAMES

Gridlock was inspired by a YouCubed task, you say? Well, I'm Jo Boaler, the founder of YouCubed. What task do you refer to? Hi, Jo, great to have you here! It's "How Close to 100?" https://www.youcubed.org/tasks/how-close-to-100.

How do I win at Tax Collector? Though there are heuristics that suffice for beating the tax collector, the optimal strategy is unknown. For more strategic ideas, check out Robert K. Moniot, "The Taxman Game," *Math Horizons*, February 2007, 18–20. By the way, thanks to math teacher Shannon Jeter and her students for suggesting the nongendered "Tax Collector" to replace the game's original name, "Taxman."

Where have I seen Starlitaire before? Perhaps in Vi Hart's timeless YouTube video "Doodling in Math Class: Stars" or Anna Weltman's *This Is Not a Math Book* (La Jolla, CA: Kane Miller, 2017).

III. COMBINATION GAMES

INTRODUCTION

Who is this Raph Koster? His book is *Theory of Fun for Game Design* (Sebastopol, CA: O'Reilly Media, 2004). For a pithy cartoon summary: Raph Koster, *A Theory of Fun: 10 Years Later* (https://www.raphkoster.com/gaming/gdco12/Koster_Raph_Theory_Fun_10.pdf). He gives his four core mechanics on page 75 of the PDF.

How do I learn more about complexity theory? I learned everything I know from my father, Jim Orlin, a leading scholar of network flows and other optimization algorithms. But I realize he is not as readily available to you. So I recommend the precipitating inspiration for my discussion in the chapter: the 99th episode of Sean Carroll's *Mindscape* podcast, "Scott Aaronson on Complexity, Computation, and Quantum Gravity."

Can you solve a Rubik's Cube? Next question.

Did the *New York Times* really cover the Fifteen Puzzle like some kind of epidemic? Yeah, but they were joking. "Fifteen," *New York Times*, March 22, 1880, page 4. At a certain point, the satire becomes clear: President Hayes comes across the puzzle (it having been placed there by "a Southern Brigadier of more than usual villainy"), and remarks, "It looks easy . . . There are fifteen numbers, and you have to arrange them so that there will be eight in one row and seven in another." Four years earlier, Hayes had won the disputed election of 1876 because a 15-member panel voted 8 to 7 in his favor. "It certainly seems to me as if I had tried that kind of puzzle somewhere," the *New York Times* imagines him saying, "though I can't at this moment recollect where it was."

SIM

Whoa, Frank Ramsey died at 27? How did he get so much done? He was 26, actually. I assume he was a filthy cheater with a time turner. Anyway, the biography of note is Cheryl Misak, *Frank Ramsey: A Sheer Excess of Powers* (Oxford, UK: Oxford University Press, 2020).

Did you say I can't memorize the winning strategy for Sim? Is that a challenge?! Hey, be my guest. The source is Ernest Mead, Alexander Rosa, and Charlotte Huang, "The Game of Sim: A Winning Strategy for the Second Player," *Mathematics Magazine* 47, no. 5 (1974): 243–247.

I am an adult who enjoys connect the dots. Can I legitimize this somewhat embarrassing hobby by learning more about Ramsey theory? Yes! Check out Martin Gardner's chapter "Sim, Chomp, and Racetrack" in *Knotted Doughnuts*. Another great introduction is Jim Propp, "Math, Games, and Ronald Graham," *Mathematical Enchantments*, July 16, 2020. Jim's writing is uniformly excellent; check out all of his monthly essays at http://mathenchant.org.

Where'd you get that nifty proof that Sim can't end in a tie? I first heard it from the excellent Yen Duong, in her appearance on the podcast *My Favorite Theorem* with hosts Evelyn Lamb and Kevin Knudson. It's episode 31. She compares the proof to broccoli with cheese sauce, as a tasty way to get kids to eat their vegetables. Too often, of course, mathematical proofs are more like uncooked radishes, taking days to chew and weeks to metabolize.

If I'm only supposed to have 150 relationships, why do I have 700 Facebook friends? Hey, I don't know your life. But for more on the anthropology, read Robin Dunbar, *How Many Friends Does One Person Need? Dunbar's Number and Other Evolutionary Quirks* (London: Faber & Faber, 2010). By the way, I learned the tale of the sociologist Sandor Szalai from Alexander Bogomolny, who quotes Noga Alon, Michael Krivelevich, and T. Gowers, eds., *The Princeton Companion to Mathematics* (Princeton, NJ: Princeton University Press, 2008), 562.

TEEKO

C'mon, you must have made up those ridiculous quotes from John Scarne, right? They come straight from John Scarne, *Scarne on Teeko* (originally published 1955; digital version published 2007 by Lybrary.com). I also drew from, and happily recommend, the funny and poignant essay by Blake Eskin, "A World of Games," *Washington Post*, July 15, 2001.

What the heck is that song whose lyrics you quote? It's Jonathan Coulton's "Skullcrusher Mountain," which in a just world would be as famous as "Purple Rain." Thanks to Jonathan for granting permission to reprint the verse.

How can I learn more about the combinatorics of language, without buying my own monkeys and typewriters? I don't even know where to buy monkeys. Or typewriters. You'll enjoy Jorge Luis Borges, "The Library of Babel," *Collected Fictions* (New York: Penguin Books, 1998). I can't remember the past, but I suspect I've cited this story in every book I've written. I also can't remember the future, but I suspect I will continue to cite it in all later books.

How did Guy Steele solve Teeko? His analysis is surprisingly hard to find; in lieu of publication, it seems to have just circulated among players. Perhaps no surprise that the best source is BoardGameGeek: https://boardgamegeek.com/thread/816476/steele-guy-november-23-1998-re-teeko-hakmem.

Can you explain the calculations in your "number of positions" chart comparing Teeko, checkers, chess, and go? Sure. An atom's mass is on the order of 10^{-23} grams. Multiply that by 7.5×10^7 (the rough number of positions in Teeko) and you get almost 10^{-15} grams, which is the rough mass of a bacterium. Multiply instead by 5×10^{20} (the rough number of positions in checkers) and you get almost 10^{-2} grams, the rough mass of a housefly. Multiply instead by 10^{41} (the rough number of positions in chess) and you get 10^{18} grams, the rough mass of Lake Huron. Multiply instead by 2×10^{170} (the rough number of positions in go), and you get 10^{147} grams, which is, uh, kind of unfathomable. It's roughly the mass you'd get if you took every atom in the visible universe and turned it into a whole visible-universe-sized object of its own.

NEIGHBORS

Where can I read more on Neighbors? The only earlier record is Sara Van Der Werf, "5x5 Most Amazing Just for Fun Game," December 13, 2015 (https://www.saravanderwerf.com/5x5-most-amazing-just-for-fun-game). I also thank Jane Kostik for sharing her memories with me.

How do you know that Neighbors came from Wordsworth? I don't, but the circumstantial evidence is pretty strong. Wordsworth is clearly older. In *Games with Pencil and Paper* (1973), Eric Solomon calls it "an old game of unknown origin" that "has been popular in England for many years," as well as "the best of all word games." In *New Rules for Classic Games* (1992), R. Wayne Schmittberger presents the same game under the name "Crossword Squares," with a few scoring changes: (1) You may count multiple words in the same row or column (as long as one is not strictly contained in the other, like *lob* in "slobs"); (2) three-, four-, and five-letter words score 10, 20, and 40, respectively; and (3) after you've created a word once, later instances on the same board only score half the point value.

Is that "Rocks Game" a real thing? It comes from Misha Glouberman and Sheila Heti, *The Chairs Are Where the People Go: How to Live, Work, and Play in the City* (New York: Farrar, Straus, and Giroux, 2011).

CORNERS

I want more Zukei puzzles. Then check out Sarah Carter's excellent blog *Math Equals Love* (https://mathequalslove.net/zukei-puzzles). Here are the solutions to the ones I showed.

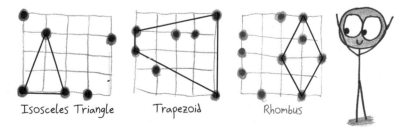

Isosceles Triangle   Trapezoid   Rhombus

What's the deal with that 17-by-17 square? I learned about it from Sam Shah, who described it in the *Aperiodical*'s 2019 "Big Internet Math-Off" (https://aperiodical.com/2019/07/the-big-internet-math-off-the-final-sameer-shah-vs-sophie-carr). I love Sam's account: "My eyes vibrate and jump around, focusing first on the blues and some of the chains they connect, when the blues fade and I immediately see all the snakes of yellows in their horizontal and vertical positions, when red randomly pops . . . I never get tired of looking at it." Anyway, it was the solution to a puzzle published by Bill Garsach, "The 17x17 Challenge. Worth $289.00. This Is Not a Joke." *Computational Complexity*, November 30, 2009. Garsach also inspired *Play With Your Math* 23, "No ReXangles" (https://playwithyourmath.com/2020/01/01/23-no-rexangles).

Wait. Did you say one of your favorite YouTube videos is a guy solving a sudoku? Yes. And not only are you about to spend the next 25 minutes watching this video, but it's going to be the highlight of your day. "The Miracle Sudoku," *Cracking the Cryptic*, May 10, 2020 (https://www.youtube.com/watch?v=yKf9aUIxdb4).

Is it spelled "Quads and Quasars" or "Quods and Quazars"? Yes, it is. Source: Ian Stewart, "Playing with Quads and Quazars," *Scientific American*, March 1, 1996, 84–85.

What else have psychologists found out about sudoku? See Hye-Sang Chang and Janet M. Gibson, "The Odd-Even Effect in Sudoku Puzzles: Effects of Working Memory, Aging, and Experience," *American Journal of Psychology* 124, no. 3 (2011): 313–324.

Where can I learn more about that chess study? William G. Chase and Herbert A. Simon, "Perception in Chess," *Cognitive Psychology* 4, no. 1 (1973): 55–81. As always in psychology, the picture is complicated. One later study suggests that experts retain some advantage (albeit a diminished one) in memorizing random boards: Fernand Gobet and Herbert A. Simon, "Recall of Rapidly Presented Random Chess Positions Is a Function of Skill," *Psychonomic Bulletin and Review* 3, no. 2 (1996): 159–163.

AMAZONS

How did you hear about this game? In *Knotted Doughnuts*, Martin Gardner discusses Quadraphage, a series of puzzles in David L. Silverman's *Your Move* (New York: McGraw-Hill, 1971). I noted down the idea, later mistook for my own, and spent months trying to make it work as a game, until an analytical sixth grader named Abby broke it. She simply ignored my king's location and lined the edge of the board with counters. (You can do even better by subdividing the board first, but the point is that Abby's insight—don't worry about the king's current location, and just build walls—reduces the game into a solvable puzzle.) I tried switching to a knight—I was essentially staggering toward Alex Randolph's game Knight Chase from *Gamut of Games*—but then another Abby, this one a computer science professor, broke that version, too. Finally someone pointed me toward Walter Zamkauskas's gem.

So you hadn't seen the Numberphile video? I hadn't. It's a great one, though. Mathematician Elwyn Berlekamp teaches the game to Numberphile host Brady Haran. Go search YouTube for "A final game with Elwyn Berlekamp (Amazons)." For more strategic detail, you can see Berlekamp's own video at his "Elwyn Berlekamp" channel. Berlekamp's death in 2019, along with those of his *Winning Ways* coauthors Richard Guy and John Conway, added a tragic note to my work on this book.

Who are all these Amazons experts you quote? They are the noble denizens of BoardGameGeek. Especially useful were the 2008 review from @cannoneer (https://boardgamegeek.com/thread/348357/why-i-love-amazons), including insightful replies such as those by Nick Bentley (@milomilo122), and the 2014 review from @ErrantDeeds (https://boardgamegeek.com/thread/1257900/amazons-walking-fine-line-between-depth-and-access). The site has lots of other great discussion of the game, too. I especially recommend David Ploog's analysis: https://www.mathematik.hu-berlin.de/~ploog/BSB/LG-Amazons.pdf.

OTHER COMBINATION GAMES

Am I really just a combination of chemicals to you? Kind of.

Where did you get that horrid idea? From the beautiful book edited by Snezana Lawrence and Mark McCartney, *Mathematicians & Their Gods: Interactions between Mathematics and Religious Beliefs* (Oxford, UK: Oxford University Press, 2015). Specifically, the chapter by Robin Wilson and John Fauvel, "The Lull before the Storm: Combinatorics in the Renaissance," from which I draw the "Sefer Yetzirah" quote.

You're saying that reducing me to a mere combinatorial exercise is a *religious* notion? A secular one, too. Just listen to Italo Calvino: "Who are we, who is each one of us, if not a combinatorial of experiences, information, books we have read, things imagined? Each life is an encyclopedia, a library, an inventory of objects . . . and everything can be constantly shuffled and reordered in every way conceivable." Italo Calvino, *Six Memos for the Next Millennium*, trans. Patrick Creagh (New York: Vintage, 1993). In Italo's telling, I am a human mishmash of Ursula Le Guin novels, Paul Simon lyrics, and Reese's Peanut Butter Cups. But that doesn't make me worthless or unoriginal; in fact, that's what originality is. Creativity is a game of combinations, of arranging the same old stuff into new and unprecedented mishmashes.

Okay, I'm slightly less offended than I was before. That's the Orlin guarantee.

IV. GAMES OF RISK AND REWARD

INTRODUCTION

In your *Deal or No Deal* example, are those real numbers? Yes. They're from the very first episode of the US version.

What was history's first treatise on probability? Drawing on the famous exchange of letters between Blaise Pascal and Pierre de Fermat, it was Christiaan Huygens, *De Rationciniis In Ludo Aleae*, 1656–1657. I accessed the translation by Richard J. Pulskamp, Department of Mathematics and Computer Science, Xavier University, July 18, 2009, https://www.cs.xu.edu /math/Sources/Huygens/sources/de%20ludo%20Aleae%20-%20rjp.pdf.

I've always wondered: Why do they play poker on *Star Trek*? It's actually a long-standing mystery in *Star Trek* fandom. "No rational person," writes game scholar Greg Costikyan, "would play roulette for the fascination of the game itself—watching a ball rolling around a wheel might be fascinating to a cat, but not to a human being."

No-stakes poker suffers the same problem. The challenge is all in the betting, and the betting is all in the stakes. So why do the senior officers of the Federation's flagship spend their free time wagering worthless chips on a random process?

To be fair, that's not the only peculiarity in the 24th century's idea of fun. Given food on demand, they do not chug milkshakes; they sip Earl Grey tea. Given a realer-than-real VR system, they never indulge in pornography, violent fantasies, or Grand Theft Auto; they cosplay works of classic literature. Maybe, when you've spent all week facing mortal risks, you crave nothing more than *meaningless* risks. Or maybe, when you're as civilized as a Starfleet officer, your idea of fun converges with a cat's.

Where can I learn more about John von Neumann? Norman Macrae, *John von Neumann: The Scientific Genius Who Pioneered the Modern Computer, Game Theory, Nuclear Deterrence, and Much More* (New York: Pantheon, 1992).

UNDERCUT

Where did you learn about this game? Douglas Hofstadter, *Metamagical Themas*. A single chapter provides the original game plus the variants Flaunt and Underwhelm.

What's that *Princess Bride* scene you're talking about? Oh my gosh, go watch the movie! *The Princess Bride*, directed by Rob Reiner, 20th Century Fox, 1987. It has this delicate, absurd touch, plus an all-time great performance from Mandy Patinkin, who sets the film's warm, self-aware tone.

How did you come up with all those examples of the value of randomness? I draw several of them (Naskapi hunters, Roman generals, bird augury, caribou bone divination) from a book review by Scott Alexander, "Book Review: The Secret of Our Success," *SlateStarCodex*, June 4, 2019. The book in question is Joseph Henrich, *The Secret of Our Success: How Culture Is Driving Human Evolution, Domesticating Our Species, and Making Us Smarter* (Princeton, NJ: Princeton University Press, 2016). The others come from years of conversations on probability and randomness with my father, the splendid Jim Orlin.

Is Morra a real game? Realer than real. Search "Morra" on YouTube for a delightful glimpse of backyard family reunions. I also recommend DW Euromaxx, "The World's Loudest Game: Morra Is the World's Oldest Hand Game—and It Is LOUD!" YouTube, August 24, 2019, https://www .youtube.com/watch?v=nEvJIG42D14.

Never mind Undercut. How do I win at rock-paper-scissors? I recommend Hannah Fry's appearance on Numberphile, hosted by Brady Haran. It's Numberphile, "Winning at Rock Paper Scissors," YouTube, January 26, 2015, https://www.youtube.com/watch?v=rudzYPHuewc. The paper she discusses is Zhijian Wang, Bin Xu, and Hai-Jun Zhou, "Social Cycling and Conditional Responses in the Rock-Paper-Scissors Game," ArXiv.org, April 21, 2014, https://arxiv.org /pdf/1404.5199v1.pdf. For my own discussion, I drew inspiration from Greg Costikyan, *Uncertainty in Games* (p. 32): "*Rocks / Paper / Scissors* is a game of player unpredictability in its purest form, for this single factor is the sole determinant of the game's uncertainty, its raison d'etre, and its cultural continuance."

Maybe you chumps have no free will. But I could defy any computer's predictions. Well, before you crow too loudly, try out the *f* or *d* predictor that Scott Aaronson described. You can find it online at https://people.ischool.berkeley.edu/~nick/aaronson-oracle, with more details here: https://github.com/elsehow/aaronson-oracle. Aaronson discusses the situation in his course Quantum Computing since Democritus. You can read more at his blog here: https://www .scottaaronson.com/blog/?p=2756.

Who came up with the three-person variant of Undercut? My elite team of sixth- and seventh-grade play testers. I believe LaRon, Abby, and Nathan devised this variant, although I don't wholly trust my memory, so I'll also thank Rohan, Allan, Charlotte, and Angela.

Is 7 really the most common "random" number? Yup. Michael Kubovy and Joseph Psotka, "The Predominance of Seven and the Apparent Spontaneity of Numerical Choices," *Journal of Experimental Psychology: Human Perception and Performance* 2, no. 2 (1976): 291–294. For a more recent example: "Asking over 8500 College Students to Pick a Random Number from 1 to 10," r/DataIsBeautiful, January 4, 2019, https://www.reddit.com/r/dataisbeautiful/comments /acow6y/asking_over_8500_students_to_pick_a_random_number.

ARPEGGIOS

Dude! Why would you bring up that study about a deadly illness, with the world still reeling from the trauma of COVID-19? I know, I know, I'm sorry. It's the classic study on the framing of risks, but I realize you probably don't want grim reality intruding on your cartoon books of games. Anyway, the original publication is Amos Tversky and Daniel Kahneman, "The Framing of Decisions and the Psychology of Choice," *Science* 211, issue 4481 (1981): 453–458. It's also discussed in Daniel Kahneman, *Thinking, Fast and Slow* (New York: Farrar, Straus, and Giroux, 2011).

Do doctors really treat age 35 as a sharp cutoff for pregnancy? Too often, yes. Read about it in Emily Oster, *Expecting Better: Why the Conventional Pregnancy Wisdom Is Wrong—and What You Really Need to Know* (New York: Penguin Books, 2013). If you're the kind of person who (a) wants or has young kids, and (b) reads the endnotes of math books, then Oster's books are 100% for you.

What are my chances of winning the single-player variant Ascender? Depends on your strategy. There is admittedly a 63% chance that the dice will prove impossible (e.g., if you roll two 2 + 2s, or three 4 + 5s). If you dislike impossible games (why would you?!), introduce a rule that you reroll if you get repeat doubles (e.g., 3-3 later followed by another 3-3) or if another roll three-peats (e.g., if you get 1-4, then another 1-4, and later another 1-4).

OUTRANGEOUS

What inspired this game? As I mention in the chapter, the impetus came from Douglas W. Hubbard, *How to Measure Anything: Finding the Value of "Intangibles" in Business* (Hoboken, NJ: John Wiley & Sons, 2007). I also thank the folks who helped to play-test it at King Edward's School, Saint Paul Academy, and Protospiel Minnesota 2019.

How did you compute those optimal strategies for the simplified version where you're just predicting a dice roll? To simplify the analysis, I assumed you must pick a range of the form 1 to n. If you allow other possibilities, the person with the narrower range wants to maximize overlap, while the person with the larger range wants to minimize it, but the basic analysis remains similar. The quantitative results also change slightly if you assume that each player rolls their own die (i.e., that the guesses are independent, which is a better model for the typical question in the game), but the basic qualitative description of the strategy remains the same.

No, I mean, how did you get the numbers? Oh! I used a handy app at the website of UCLA Professor Thomas Ferguson: https://www.math.ucla.edu/~tom/gamesolve.html.

What kind of fool claims to be 100% confident on anything? Just your garden-variety human fool. For one example, see Pauline Austin Adams and Joe K. Adams, "Confidence in the Recognition and Reproduction of Words Difficult to Spell," *American Journal of Psychology* 73, no. 4 (1960): 544–552. Another good source on these issues is Daniel Kahneman's book *Thinking, Fast and Slow*.

Where should I go to learn more about good calibration? My go-to writer on these kind of epistemic issues is Julia Galef. Her engaging and thoughtful book is: Julia Galef, *The Scout Mindset: Why Some People See Things Clearly and Others Don't* (New York: Portfolio, 2021).

PAPER BOXING

Where can I learn more about the politics of gerrymandering? I recommend David Litt, *Democracy in One Book or Less: How It Works, Why It Doesn't, and Why Fixing It Is Easier than You Think* (New York: Ecco, 2020).

Where can I learn more about the mathematics of gerrymandering? I recommend the work of Moon Duchin. She appeared on the *Quanta* podcast *Joy of X*, hosted by Steven Strogatz, in the episode "Moon Duchin on Fair Voting and Random Walks." Her research team at Tufts University is the Metric Geometry and Gerrymandering Group, http://mggg.org.

Where can I learn more about salamanders that look like Elbridge Gerry? I fear you may have misunderstood what gerrymandering is.

Who made that wild golf metaphor? Zach McArthur, a high school math teacher and golf coach in Chicago. Thanks to Michael Hurley for putting us in touch.

RACETRACK

Where did you learn about this game? My original source was Martin Gardner, *Knotted Doughnuts*. I drew some variants from Andrea Angiolino, *Super Sharp Pencil and Paper Games*.

Do you believe that the future is undecided ("uncertain" in the ontological sense), or merely unrevealed ("uncertain" in the epistemic sense)? Undecided.

You haven't decided? No, I mean the future is undecided. The uncertainty is ontological.

Oh, great! So we do have free will? Not at all. The uncertainty resides at the quantum level, and then propagates to larger scales. There's no role for conscious human will.

So . . . we *don't* have free will? No, but it's a useful fiction, so don't worry too much about it.

You just told me I have no free will! How can I not worry about it? Well, look at it this way. You should only worry about what you can control, and in a world without free will, you control nothing. Thus, you should worry about nothing. Problem solved!

OTHER GAMES

Where do I find that rock-paper-scissors variant with 101 different gestures? Its creator is David C. Lovelace, and it's called "RPS 101: The Most Terrifying Complex Game Ever." You can find it online: https://www.umop.com/rps101.htm.

Why do we need more gestures in rock-paper-scissors, anyway? In his description of the Lizard-Spock variant, Sam Kass muses that "it seems like when you know someone well enough, 75–80% of any Rock-Paper-Scissors games you play with that person end up in a tie." The writers of *The Big Bang Theory* cite this made-up number as if it's some kind of empirical fact, which is why you shouldn't get your social science lectures from sitcoms. At least not CBS sitcoms. http:// www.samkass.com/theories/RPSSL.html.

Where did you get 101 and You're Done? Marilyn Burns, *About Teaching Mathematics* (Sausalito, CA: Math Solutions Publications, 2007). She calls it 101, You're Out; I thank Robert Biemesderfer for the suggestion of the rhyming alternative 101 and You're Done.

How do I win at 101 and You're Done? Here's a simple improvement on the greedy algorithm. Assign a predicted value to each remaining die, and only multiply by 10 if the remaining dice will not put you over the top. A predicted value of 0 gives the greedy algorithm; effectively, you're pretending later dice don't exist. A predicted value of 6 gives a super-safe algorithm, with no risk of ever going over. In short, the higher your predicted value, the more cautious your strategy. In simulations, I found that a predicted value of 4.5 gave the highest average score: 88.4 points per round, going bust only 1.5% of the time.

Have you ever played the Con Game? No, but its creator, James Ernest, has. See James Ernest, *Chief Herman's Holiday Fun Pack: Instruction Booklet and Guide to Better Living*.

Where did Breaking Rank come from? A conversation with my dad, Jim Orlin, in which we sought to modify and improve Outrangeous. I still prefer Outrangeous (it's far easier to come up with good questions), but I find the scoring of Breaking Rank more intuitive and elegant.

How do I win at Pig? First, here's a flawed approach: Commit to a certain number of rolls each turn. "I shall roll *n* times, then stop, no matter what my score is." A little calculus turns up the optimal value of $1 / (\ln 18 - \ln 13)$, or roughly 3.07 rolls. In other words, roll three times, then quit. This delivers about 11.5 points on average.

The thing is, who cares how many times you've rolled? What matters is your score. A better rule takes the form "I shall roll until I reach score *x*, then stop, no matter how many rolls it takes me." A little probability theory reveals the optimal number is 26.5. Thus, with 26 points or fewer, keep rolling. With 27 or more, bank them. This strategy outperforms the "roll three times" strategy by about 0.4 points per turn.

V. INFORMATION GAMES

INTRODUCTION

I desire more information on "information." You can read the historic paper itself: Claude Shannon, "A Mathematical Theory of Communication" (two parts), *Bell System Technical Journal* 27, no. 3 (1948): 379–423, and 27, no. 4 (1948): 623–656. For this chapter, I also relied on two great secondary sources: James Gleick, *The Information: A History, a Theory, a Flood* (New York: First Vintage Books, 2011), and Jimmy Soni and Rob Goodman, *A Mind at Play: How Claude Shannon Invented the Information Age* (New York: Simon and Schuster, 2017).

BULLSEYES AND CLOSE CALLS

What's the optimal strategy? One sensible approach is a *minimax* algorithm. For each possible guess, imagine the most disappointing feedback possible—that is, the feedback that would allow you to cross off the fewest options. Then, of all these worst-case scenarios, find the most informative. You should guess that number, whose minimal feedback is a maximum. Hence "minimax." For more, check out Donald E. Knuth, "The Computer as Master Mind," *Journal of Recreational Mathematics* 9, no. 1 (1976–1977): 1–6.

Hey, I never noticed the conceptual connection between "probe" and "problem." The insight comes from Paul Lockhart, *Measurement* (Cambridge, MA: Belknap Press, 2014).

I don't like that card-flipping study. You tricked me. Well, if you want to mount a class-action lawsuit, 96% of subjects can join you. Anyway, it's called the Wason selection task; see P. C. Wason and Diana Shapiro, "Natural and Contrived Experience in a Reasoning Problem," *Quarterly Journal of Experimental Psychology* 23 (1971): 63–71. Also, if you want to get way better at the task, just turn the letters into beverages, turn the numbers into ages, and enforce the rule "Any card with an alcoholic drink must have an age over 21." It's the same logical structure, but a much easier problem.

CAVEAT EMPTOR

Are winners truly cursed? I mean, it's nothing supernatural, like the curse of appearing on the *Sports Illustrated* cover, but it's just as real. For a quick overview, check out Adam Hayes, "Winner's Curse," *Investopedia*, November 8, 2019.

Can you balance out this "curse" talk with an uplifting tale of how good humans are at estimating? Sure. Go read James Surowiecki, *The Wisdom of Crowds* (New York: Anchor, 2005), from which I draw the example of estimating the ox's weight.

Thanks. I'm ready to be devastated again by the depravity of mankind. Then go read Michael Lewis, *Liar's Poker: Rising through the Wreckage on Wall Street* (New York: W. W. Norton, 1989), from which I draw the game of the title.

Ugh. That hurt my soul. Okay, cheer yourself up with the schadenfreude of David D. Kirkpatrick, "Mystery Buyer of $450 Million 'Salvator Mundi' Was a Saudi Prince," *New York Times*, December 6, 2017. The seller attributed the painting to Leonardo da Vinci, but scholars have been disputing this claim. Convincingly, if you ask me. (The Saudi prince did not ask me.) Anyway, imagine paying half a billion dollars for a Leonardo painting that isn't by Leonardo.

Who helped you create this game? Big shout-out to play-testing pals Matt Donald, Rob Liebhart, and Jeff Bye.

LAP

Where did you learn about this game? *Gamut of Games.* But my real education came from a great thread on BoardGameGeek, where users @russ, @LarryLevy, @mathgrant, and @Bart119 explored strategic ideas and noted the existence of the ambiguous board design: https://boardgamegeek.com/thread/712697. Meanwhile, the neat Rainbow Logic variant comes from Elizabeth Cohen and Rachel Lotan, *Designing Groupwork: Strategies for the Heterogeneous Classroom* (New York: Teachers College Press, 2014).

QUANTUM GO FISH

Where did you hear about this game? In hindsight, someone tried to teach it to me during my wife's grad program at UC Berkeley. But only while researching this book did I come across Anton Geraschenko's account: http://stacky.net/wiki/index.php?title=Quantum_Go_Fish. The description at Everything2 is also good: https://everything2.com/title/Quantum+Fingers. And it's worth checking out the r/math discussion "Quantum Go Fish (a True Mathematicians' Card Game)" on Reddit.

Is it weird that this reminds me of Negative Twenty Questions? I had the same thought.

Physicist John Wheeler used this game to illustrate how the questions we choose to ask shape our view of reality. You send one person, the guesser, out of the room. Everyone else then agrees not on an object, but merely on a specific pattern of yes and no answers—for example, yes, yes, no, yes, yes, no, and repeating that way forever. Always answer this way, no matter what the guesser asks (except if it would contradict an earlier answer). The guesser's questions will, on their own, create the object they are asking about. Whatever the guesser settles on is the "right" answer.

Here's a sample round, using the pattern yes/yes/no. Is it alive? Yes. Is it a person? Yes. Are they male? No. Are they female? Yes. Is she famous? Yes. Is she an entertainer? No. Is she a political leader? Yes. Is she a current or former head of state? Yes. Is her first language English? No. Is she European? Yes. Is it Angela Merkel? Yes.

How good were you at Rush Hour? Pretty decent, though I don't have a set anymore. It's manufactured by ThinkFun these days, if you want to send me a gift.

SAESARA

Where can I learn more about inductive games? You can begin with Martin Gardner, *Origami, Eleusis, and the Soma Cube* (New York: Cambridge University Press, 2008). For the thoughts of Eleusis's inventor, check out Robert Abbott, "Eleusis and Eleusis Express," LogicMazes.com, http://www.logicmazes.com/games/eleusis. I also benefited greatly from the detailed "Zendo—Design History" by the game's creator Kory Heath, at his personal blog (http://www.koryheath.com/zendo/design-history).

I think you're overstating the danger of gathering data without a clear hypothesis. I'm not. For details, read my chapter on p-hacking and the replication crisis: "Barbarians at the Gate of Science," in Ben Orlin, *Math with Bad Drawings: Illuminating the Ideas That Shape Our Reality* (New York: Black Dog & Leventhal, 2018).

Did Darwin actually say "I hate everybody & everything"? Yes. Robert Krulwich, "Charles Darwin and the Terrible, Horrible, No Good, Very Bad Day," NPR, *Krulwich Wonders* blog, October 19, 2012, https://www.npr.org/sections/krulwich/2012/10/18/163181524/charles-darwin-and-the-terrible-horrible-no-good-very-bad-day.

Did Einstein actually say that he "seldom" has ideas? Yes. Bill Bryson, *A Short History of Nearly Everything* (New York: Broadway Books, 2003).

Did R. C. Buck actually say that "creativity is the heart and soul of mathematics"? I understand your challenging me on the Darwin and Einstein quotes, which sound like they might be jokes, but you're really going to ask for a source on this one?

Yes. Okay. I got it from Denise Gaskins, "Quotations XV: More Joy of Mathematics," *Denise Gaskins' Let's Play Math*, https://denisegaskins.com/2007/09/19/quotations-xv-more-joy-of-mathematics. She got it from John A. Brown and John R. Mayor, "Teaching Machines and Mathematics Programs," *American Mathematical Monthly* 69, no. 6 (1962): 552–565.

OTHER INFORMATION GAMES

Is Win-Lose-Banana really rich and strategic? Or just frivolous nonsense? That's for you to decide, although one ingenious Redditor (u/tdhsmith) argues that by asking this question, "you've just entered the game of *Meta*-Win-Lose-Banana. The game is thematically similar, but played out in the real world. There are three factions:

"The uncertain (win) faction . . . aren't sure whether *Win, Lose, Banana* is strategic or not. This is your current faction. When you join this faction, you reveal immediately by asking whether the game has strategy, which triggers the second phase of the game.

"The frivolous (lose) faction . . . are trying to convince the uncertains that the game has no strategy. If the uncertain faction sides with them, only the frivolous win, because they had the superior viewpoint and the uncertains wasted their time on the game.

"The strategic (banana) faction . . . are trying to convince the uncertains that the game is rich and strategic. If the uncertain faction sides with them, they both win, because they had a fun and strategic time."

By the way, I heard about the game from Jonny Bouthilet and Marcus Ross. It was designed by Chris Cieslik, with art by Cara Judd, and published by Asmadi Games.

INDEX

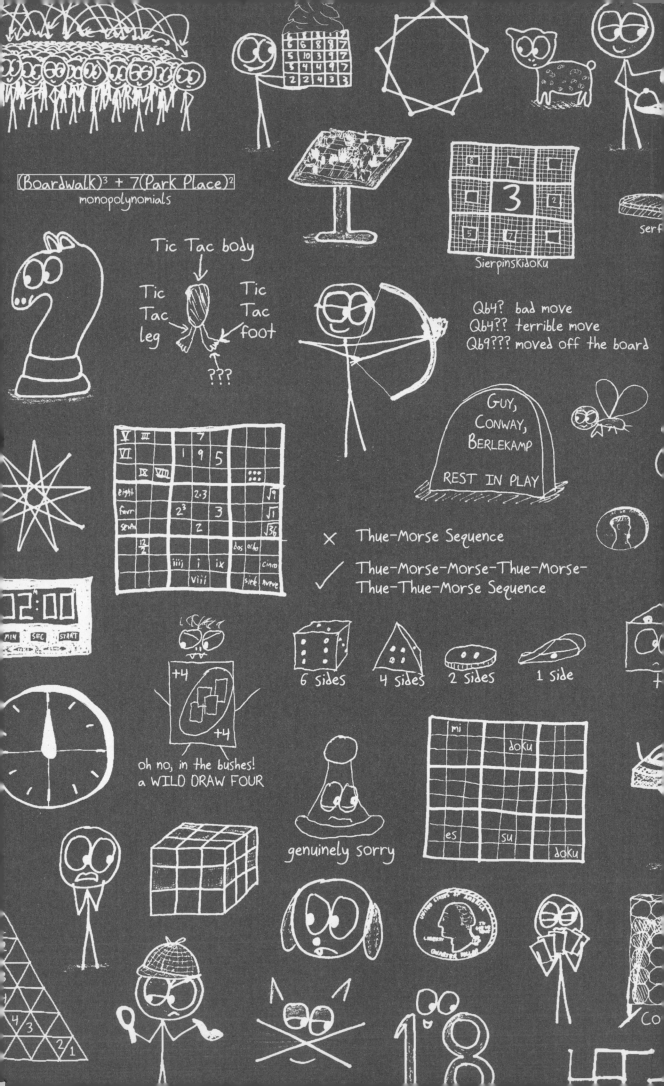